Lecture Notes
in Business Information Processing 162

Series Editors

Wil van der Aalst
  Eindhoven Technical University, The Netherlands
John Mylopoulos
  University of Trento, Italy
Michael Rosemann
  Queensland University of Technology, Brisbane, Qld, Australia
Michael J. Shaw
  University of Illinois, Urbana-Champaign, IL, USA
Clemens Szyperski
  Microsoft Research, Redmond, WA, USA

T0224051

Philippe Cudre-Mauroux
Paolo Ceravolo
Dragan Gašević (Eds.)

Data-Driven Process Discovery and Analysis

Second IFIP WG 2.6, 2.12 International Symposium
SIMPDA 2012, Campione d'Italia, Italy, June 18-20, 2012
Revised Selected Papers

 Springer

Volume Editors

Philippe Cudre-Mauroux
University of Fribourg, Switzerland
E-mail: phil@exascale.info

Paolo Ceravolo
Università degli Studi di Milano, Italy
E-mail: paolo.ceravolo@unimi.it

Dragan Gašević
Athabasca University, AB, Canada
E-mail: dgasevic@acm.org

ISSN 1865-1348   e-ISSN 1865-1356
ISBN 978-3-642-40918-9   e-ISBN 978-3-642-40919-6
DOI 10.1007/978-3-642-40919-6
Springer Heidelberg New York Dordrecht London

Library of Congress Control Number: 2013947989

Typesetting: Camera-ready by author, data conversion by Scientific Publishing Services, Chennai, India

Printed on acid-free paper

Springer is part of Springer Science+Business Media (www.springer.com)

Preface

The rapid growth of organizational and business processes managed via information systems made a big variety of data available that consequently created a high demand for making available data analysis techniques more effective and valuable. The second edition of the International Symposium on Data-Driven Process Discovery and Analysis (SIMPDA 2012) was conceived to offer a forum where researchers from different communities and industry can share their insights in this hot new field. The symposium featured a number of advanced keynotes illustrating new approaches, presentations on recent research, a competitive PhD seminar, and selected research and industrial demonstrations. The goal is to foster exchanges among academic researchers, industry, and a wider audience interested in process discovery and analysis. The event was organized jointly by the IFIP WG 2.6 and W.G 2.12.

Submissions covered theoretical issues related to process representation, discovery, and analysis or provided practical and operational experiences in process discovery and analysis. To improve the quality of the contributions the symposium fostered the discussion during the presentation, giving authors the opportunity to improve their work by extending the presented results. For this reason, authors of accepted papers and keynote speakers were invited to submit extended articles to this post-symposium volume of LNBIP. There 17 submissions and 6 papers accepted for publication.

In the first paper "Lightweight RDF Data Model for Business Processes Analysis," Marcello Leida et al., presents a lightweight data representation model to implement business process monitoring transparently to the data creation process.

The second paper by Santiago Aguirre et al., "Combination of Process Mining and Simulation Techniques for Business Process Redesign: a Methodology Approach," addresses the problem of using simulation for organizing process redesign. In particular the simulation model is constructed based on the discovery analysis and on the waiting times calculated through a statistical analysis of the event log data.

The third paper by Wil van der Aalst et al., "Improving Business Process Models Using Observed Behavior," proposes a technique for aligning reference process models to observed behaviours, introducing five quality dimensions that are balanced and used to introduce enrich the reference model. In particular this work analyses the effect of introducing the similarity dimension in addition to other dimensions previously adopted in the literature.

The fourth paper by Sjoerd van der Spoel et al., "Process Prediction in Noisy Data Sets: A Case Study in a Dutch Hospital," applies classifier algorithms for predicting the outcome and duration of a process with the objective of improving the capability of predicting the cash flow of health care organizations.

The fifth paper by Andreas Wombacher et al., "Towards Automatic Capturing of Semi-Structured Process Provenance", discusses the problem of integrating provenance information with process monitoring systems, improving the automation of the procedure. The authors propose to achieve this result by using the access logs on file used in the execution of the process and provide a demonstration of the analysis that can be performed using that approach.

The sixth paper by Jan Mendling, "Managing Structural and Textual Quality of Business Process Modelss," gives an overview on how empirical research informs structural and textual quality assurance of process models.

We gratefully acknowledge the strong research community that gathered around the research problems related to process data analysis and the high quality of their research work, which is hopefully reflected in the papers of this issue. We would also like to express our deep appreciation for the referees' hard work and dedication. Above all, thanks are due to the authors for submitting the best results of their work to the symposium on Data-Driven Process Discovery and Analysis.

We are very grateful to the Università degli Studi di Milano and to IFIP for their financial support, and to the University of Fribourg, and the University of Athabasca.

August 2013

Paolo Ceravolo
Philippe Cudre-Mauroux
Dragan Gasevic

Organization

Conference Co-Chairs

Paolo Ceravolo   Università degli Studi di Milano, Italy
Philippe Cudre-Mauroux   University of Fribourg, Switzerland
Dragan Gasevic   Athabasca University, Canada

Advisory Board

Tharam Dillon   DEBII, Curtin University, Australia
Ernesto Damiani   Università degli Studi di Milano, Italy
Moataz A. Ahmed   KFUPM, Saudi Arabia
Elizabeth Chang   DEBII, Curtin University, Australia
Erich Neuhold   University of Vienna, Austria
Karl Aberer   EPFL, Switzerland

Demonstration and Showcase Committee

Elizabeth Chang   DEBII, Curtin University, Australia
Christian Guetl   Graz University of Technology, Austria

Publicity Chair

Matthew Smith   Leibniz University Hannover, Germany

PhD. Award Committee

Gregorio Piccoli   Zucchetti spa, Italy
Paolo Ceravolo   Università degli Studi di Milano, Italy
Farookh Hussain   Curtin University, Australia

Web Chair

Fulvio Frati   Università degli Studi di Milano, Italy

Program Committee

Peter Spyns   Free University of Brussels, Belgium
Rafael Accorsi   University of Freiburg, Germany

Irene Vanderfeesten	Eindhoven University of Technology, The Netherlands
Daniele Bonetta	Università della Svizzera Italiana, Swizerland
Etienne Rivière	Université de Neuchatel, Swizerland
Sylvain Hallé	Université du Québec à Chicoutimi, Canada
Ioana Georgiana	Free University of Brussels, Belgium
Schahram Dustdar	Vienna University of Technology, Austria
Hong-Linh Truong	Vienna University of Technology, Austria
Ebrahim Bagheri	Athabasca University, Canada
Mustafa Jarrar	Birzeit University, Palestinian Territory
Hamid Motahari	HP Labs, USA
Valentina Emilia	Balas, University of Arad, Romania
George Spanoudakis	City University London, UK
Gregorio Martinez Perez	University of Murcia, Spain
Wei-Chiang Hong	Oriental Institute of Technology, Taiwan (China)
Mohamed Mosbah	University of Bordeaux, France
Jerzy Korczak	Wroclaw University of Economics, Poland
Jan Mendling	Wirtschaftsuniversitat Wien, Austria
Maurice van Keulen	University of Twente, The Netherlands
Bernhard Bauer	University of Augsburg, Germany
Christos Kloukinas	City University London, UK
Gabriele Ruffatti	Engineering Group, Italy
Alessandra Toninelli	Engineering Group, Italy
Eduardo Fernandez-Medina	University of Castilla-La Mancha, Spain
Chi Hung	Tsinghua University, China
Nora Cuppens	Telecom Bretagne, France
Debasis Giri	Haldia Institute of Technology, India
Wil Van der Aalst	Technische Universiteit Eindhoven, The Netherlands
Antonio Mana Gomez	University of Malaga, Spain
Davide Storelli	Università del Salento, Italy
Jose M. Alcaraz Calero	Hewlett-Packard Labs, UK

Table of Contents

A Lightweight RDF Data Model for Business Process Analysis

Marcello Leida, Basim Majeed, Maurizio Colombo, and Andrej Chu

EBTIC (Etisalat BT Innovation Center), Khalifa University, P.O. Box 127788,
Abu Dhabi, U.A.E.

Abstract. This article presents a lightweight data representation model
designed to support real time monitoring of business processes. The
model is based on a shared vocabulary defined using open standard
representations (RDF) allowing independence and extremely flexible in-
teroperability between applications. The main benefit of this representation is that it is transparent to the data creation and analysis processes;
furthermore it can be extended progressively when new information is
available. Business Process data represented with this data model can
be easily published on-line and shared between applications. After the
definition of the data model, in this article, we demonstrate that with the
use of this representation it is possible to retrieve and make use of domain specific information without any previous knowledge of the process.
This model is a novel approach to real-time process data representation
and paves the road to a complete new breed of applications for business
process analysis.

Keywords: business process, ontology, RDF, SPARQL, linked data,
real time processing.

1 Introduction

Business Process Management (BPM) is increasingly taking a place as a critical
area of information technology, helping businesses to leverage their resources for
maximum benefit. It has been successfully applied at improving the performance
of business processes:

- by automatically extracting process models from existing systems logs, event
  logs, database transactions, audit trail events or other sources of information
  [1,2];
- by allowing instant analysis of the business processes using interactive visual
  displays of the process workflow;
- by identifying specific case characteristics or trends that influence processing
  times [3].

However, existing BPM systems tend to limit the kind of data they can analyse
and rigid on the data model itself. Moreover, most systems offer very limited
real-time analysis capabilities. Furthermore, the kind of analysis that can be

P. Cudre-Mauroux, P. Ceravolo, and D. Gašević (Eds.): SIMPDA 2012, LNBIP 162, pp. 1–23, 2013.

performed is normally restricted to an inextensible and inflexible set of functions tied to the underlying model, besides others challenges identified by the IEEE Task Force on Process Mining [4].

The current technology typically requires the translation or import of data into the system before it can be analysed [1,2,3]. This imposes severe constraints on the way the data is represented and collected. The main limitation is however imposed by schemas that are neither flexible, nor easily extensible, meaning that new information cannot be easily captured by the system without off line modifications; moreover the high complexity of the model used makes such modification expensive. The scope for analysis and information extraction is hence severely limited.

In order to address these issues, in this article we present a lightweight and flexible business process representation model. Lightweight in the sense that the model presented is a very simple conceptualisation of a business process, that can be extended by domain specific conceptualisations. The model we present in this paper is an extremely reduced set of concepts allowing third party client to perform independent process analysis.

The approach is based on the Resource Description Framework (RDF) [5] standard that allows independence between the applications generating business process data and those consuming it. The approach is validated by a proof of concept prototype based on a set of SPARQL [6] queries that demonstrates that the applications can independently work without the need for exchanging domain specific information about the business process monitored.

The article is structured as follows: Section 2 introduces the actual state of the art; Section 3, the main section of the article, describes the basic data model used in our system and a use case scenario where the basic data model is extended by a process monitoring application with domain specific information. Then, Section 4 demonstrates how an analytical tool can perform specific analysis without previous knowledge of the process by using a set of SPARQL queries. Section 5 focuses on how the schema can be used for real time monitoring applications. Section 6 outlines open challenges that the use of such model can introduce. Finally, Section 7 presents final consideration and future work on this area.

2   Related Work

There is a very broad literature and commercial production related to business processes and their representation. However, in this section we will limit the scope to the data models that have been defined starting from requirements such as flexibility, extensibility and suitability for real-time analysis.

There are well known languages, such as BPMN[1] and BPEL[2] that are used to explicitly define a process and capture execution information for fully automated systems (e.g. web service orchestration). However in most cases integration of

---

[1] http://www.omg.org/spec/BPMN/2.0/
[2] http://docs.oasis-open.org/wsbpel/2.0/OS/wsbpel-v2.0-OS.html

BPEL engines in existing and on-going processes is a non-trivial undertaking and it is often an effort that enterprises prefer to avoid unless a minimum Return of Investment (ROI) is guaranteed. This effort might also be undesirable when the process is not formally captured and exists only in an idealized sense in the mind of the people involved; this is particularly the case for many SMEs.

Therefore, there are many situations where a process model is of very limited use or simply not available: in this case, the process model needs to be inferred from the information generated during its execution, following a bottom-up approach to build the process model from execution data. This is the case of many business process mining tools, such as Aperture [3], ProM [1] and Aris[7].

The data model underpinning Aperture is based on a flexible database model based on the concept ofprocess and task instances, with arbitrary number and type of attributes.

The MXML data format [2] and the Extensible Event Stream (XES) standard[3] used by ProM and Aris are, as for the Aperture model, general purpose, flexible and extensible business process representations.

However, the main issue with this approach is that they are strongly oriented toward a task-based representation, thus it is extremely difficult to analyse the flow from a different point of view. Moreover these models are based on a tree representation, while our model is based on a graph, which is expressively more powerful.

Ontology-based business process representation has also been defined: in [8], the authors define an ontology for semantically annotate business processes: the work focuses on the web service domain and the outcome is a set of ontologies and tools used to analyse the processes. The ontologies clearly describes the processes, the actors (also as organizations) involved, and all the tasks and constraints composing the processes. The target of this work is to use these ontologies to validate process executions by consistency checking on the processes represented as instances of the ontology.

Another example of business process models defined as ontologies is Web Service Modeling Ontology (WSMO) [9], which has been applied in the field of Web Service composition and execution. The approach is very similar to [8] and provides a set of ontologies to be used for business process validation.

The major issue with both approaches is that the process model is already defined by the ontology and the data is usually related to the concept by a mapping. This is an approach that is valid in case of fully automated processes (such as in a Web Service scenario), processes that can be easily deployed and that generate clean data, because these approaches require the data to be translated into the ontological format (if not generated already in the ontological format).

In our approach we do not define a process-specific ontology, rather the ontology is created dynamically starting from the data generated by process executions. Hence we can potentially represent any process without having to make any change to the data model. This particular situation is the one we plan to

---

[3] http://www.xes-standard.org/

challenge with the solution presented in this article: the data model presented is based on the Resource Description Framework (RDF) and is more naturally suited to represent continuously evolving and unpredictable knowledge associated with business processes.

3   An Extensible Business Process Data Model

As already reported in Section 2, the main limitation in actual business process analysis tools is the tight connection between the applications that capture the information about the process and the ones that analyse it, mainly due to the rigid data model used. Nowadays we face the need to deal with increasingly complex systems and monitoring tools; that have to be flexible and robust enough to process also information that is not present or unknown at the moment of deployment or system initialization. Next generation Business Intelligence (BI) systems will need to react to real-time events and predict the process behaviour in order to take corrective actions [10]. Moreover it is necessary to separate the data creation step from the analytical one in terms of meta-data dependency: departments that execute the processes are mostly unaware of the type of analysis that will be performed and they tend not to communicate small changes in the way and quantity of data they capture [3]. On the other side, also the analytical applications do not need to know how the information is captured, as long as it is made available on time and is compliant to a standard representation. Therefore, the need for a less constraining and less rigid model is arising. To this aim, it is important to use an extensible, flexible and publicly available data model. We identified the Resource Description Framework (RDF) [5] and RDF Schema (RDFS) [11] as a very elegant way to solve this issue.

RDF is a standard vocabulary definition which is at the basis of the Semantic Web vision, it is composed of three elements: *concepts*, *relations* between concepts and *attributes* of concepts. These elements are modelled as a labelled oriented graph [12], defined by a set of triples <s,p,o> where s is *subject*, p is *predicate* and o is *object*. Formally a graph G can be defined as:

$$G \equiv (U \cup B \cup L) \times U \times (U \cup B \cup L)$$

where:

- U is an infinite set of constant values (called URI references) these have their well-defined semantics provided as an example by the RDF and RDFS vocabularies;
- B is an infinite set of identifiers (called Blank nodes) which identify instantiation of concepts. Elements in this set do not have a defined semantic;
- L is an infinite set of values (called Literals). Elements in this set do not have a defined semantic.

The elements of a triple <s,p,o> are respectively: $s \in (U \cup B \cup L)$, $p \in U$, and $o \in (U \cup B \cup L)$. The elements in U represents elements of the schema as in Figure 1, while the elements in B and L are used to represent instances.

An RDF graph is defined by a list of triples and adding new information reduces to append new triples to the list. It is therefore easy to understand why such representation can provide big benefits for real time business process analysis: data can be appended on the fly to the existing one and it will become part of the graph, available for any analytical application, without need for reconfiguration or any other data preparation steps as we will see in Section 5.

RDF is an extremely generic data representation model and it can be used in any domain. The level of flexibility and extensibility offered by the RDF model however requires high computational power in order to query the information in the graph. A system which is able to store and query an RDF graph is called a triple store. It is out of the scope of this article to go in details of the features of a triple store; in order to get a detailed overview the reader should refer to [13] [14] for some comparative surveys. In order to get some comparative measures for query answering times of the main triples stores please refer to [15].

RDF and RDFS standard vocabularies allow external application to query data through SPARQL query language [6]. SPARQL is a standard query language for RDF graphs based on conjunctive queries on triple patterns, which identify paths in the RDF graph. SPARQL is supported by most of the triples stores available.

RDF and RDFS provide a basic set of semantics that is used to define concepts, sub-concepts, relations, attributes, and can be extended easily with any domain-specific information. In the specific case of this article, the application domain is related to Business Processes analysis and in order to provide a point of contact for the data generation and analytical applications, it is important to agree on a common vocabulary defining the set of semantics of this domain: a simple representation allowing a high level of flexibility and at the same time providing a set of elements (concepts, relations and attributes) that are meaningful in the business process analysis domain. We identified as the most generic business process the conceptual model represented in Figure 1.

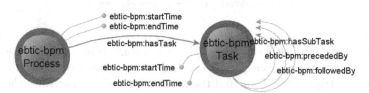

Fig. 1. The conceptual model representing the basic business process representation

The main elements are the concepts: *ebtic-bpm:Process* that represents the business process and *ebtic-bpm:Task*, representing the activity that composes the processthe process. They both have a basic set of attributes representing the beginning and termination of an activity: *ebtic-bpm:startTime* and *ebtic-bpm:endTime*. A relation between these two basic concepts defines the simplest way to represent the workflow of a business process execution: *ebtic-bpm:hasTask* is defined from the *ebtic-bpm:Process* to the *ebtic-bpm:Task* concepts and represents the set of tasks belonging to a process. Finally, a set of recursive relations are defined on the *ebtic-bpm:Task* concept: *ebtic-bpm:followedBy*, *ebtic-bpm:precededBy* and *ebtic-bpm:hasSubTask*. They respectively indicate which tasks precede and follow a given one and which tasks are subtasks of a given one. The *ebtic-bpm:followedBy* and *ebtic-bpm:precededBy* relations are used to build the process workflow while the *ebtic-bpm:hasSubTask* is useful in order to define stages of execution and to ease the analysis and structural organisation of complex processes. It is important also to underline the fact that the representation of Figure 1 does not contain any data, it provides only a conceptual schema that the process monitoring applications will extend and instantiate as described in Section 3.1.

The conceptual model of Figure 1 can be formalised by a graph G defined by the set of triples in Table 1[4]. The elements in the conceptual model of Figure 1 are defined as a vocabulary EBTIC-BPM that extends the set U which already contains the vocabularies RDF and RDFS.

Table 1. The list of triples that defines the basic Business process representation with elements from EBTIC-BPM vocabulary

s	p	o
ebtic-bpm:hasTask	rdfs:range	ebtic-bpm:Task
ebtic-bpm:hasTask	rdfs:domain	ebtic-bpm:Process
ebtic-bpm:precededBy	rdfs:domain	ebtic-bpm:Task
ebtic-bpm:precededBy	rdfs:range	ebtic-bpm:Task
ebtic-bpm:followedBy	rdfs:domain	ebtic-bpm:Task
ebtic-bpm:followedBy	rdfs:range	ebtic-bpm:Task
ebtic-bpm:hasSubTask	rdfs:domain	ebtic-bpm:Task
ebtic-bpm:hasSubTask	rdfs:range	ebtic-bpm:Task
ebtic-bpm:endTime	rdfs:range	xs:dateTime
ebtic-bpm:endTime	rdfs:domain	ebtic-bpm:Process
ebtic-bpm:endTime	rdfs:domain	ebtic-bpm:Task
ebtic-bpm:startTime	rdfs:range	xs:dateTime
ebtic-bpm:startTime	rdfs:domain	ebtic-bpm:Task
ebtic-bpm:startTime	rdfs:domain	ebtic-bpm:Process

The RDF graph resulting by these triples is displayed in Figure 2: the nodes represents the elements from the $(U \cup B \cup L)$, the number of the connections represents the triple identifiers, and the label of connection identifies the triple part: the label SP is the part of the triple connecting *subject* to *object* and the PO is the part of the triple connecting *predicate* to *object*.

According to **Linked Data** [16] principles the ebtic-bpm RDF graph will be made available on-line and advertised as a publicly available schema. So that business process data can be published and shared using this representation. This will allow consumer application to be able to process any business process defined according to this extensible schema, as described in Section 4.

---

[4] The elements with xs: preamble are part of the XML Schema vocabulary (http://www.w3.org/XML/Schema).

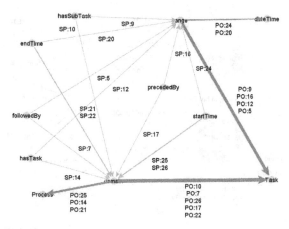

Fig. 2. The RDF graph resulting by the triples in Table 1

3.1   Extending the EBTIC-BPM Vocabulary

The conceptual model defined by the triples in Table 1 is generic and does not provide any domain specific information that can be used for a meaningful business process analysis; this information needs to be provided independently by the applications that monitor the process execution and defined as an extension of the EBTIC-BPM vocabulary. We will not go into the details of this application since it depends on the specific deployment configuration, as an example this application could be deployed on a SOA environment and monitor the message exchange on an Enterprise Service Bus (ESB), it could monitor a set of log files, it could be defined as a trigger in a database and so on. What is relevant is that the information captured by this application has to be interpreted and translated into triples.

The process monitor will act as a bridge between the specific execution environment of the process to monitor and the triple store where the triples will be directed to; in order to make accessible the information to other applications through the SPARQL interface of the triples store as described more in detail in section 3.2.

As an example let us consider the process of building a product A; this product is composed by a set of components that need to be assembled. Once the product is assembled, its correct behaviour is verified by a testing activity. It is possible to capture the semantic of this process by defining a vocabulary Product-A (PA) that enriches the basic process model EBTIC-BPM as represented by the conceptual model in Figure 3. This vocabulary is not a rigid process model definition such as BPML but just the list of attributes, relations and tasks that *may* appear during the execution of the process.

Modelling the information as an RDF graph allows monitoring applications to provide also an incomplete vocabulary or even not to provide any vocabulary at all during start time, and update it when a new concept, task or an attribute appears during the execution of the process; hence also the process definition

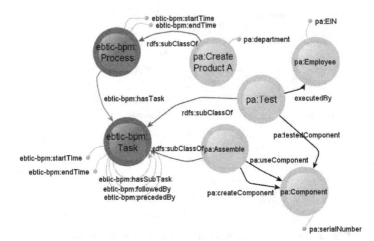

Fig. 3. The EBTIC-BPM conceptual model extended with domain specific concepts

is provided on-line. This is an important assumption that allows our approach to adapt to situations where the process is not known a priori and where its behaviour is not predictable or cannot be bounded. Updating the vocabulary in this model is equivalent to appending the related triples to the graph and there is no constraint of any type on when these triples should appear in the graph. As it is also possible to notice from Figure 4, this action is transalted in adding nodes and relations to the graph.

The domain specific vocabulary is composed by a set of concepts, relations and attributes:

- *pa:CreateProductA*: this concept extends the concept *ebtic-bpm:Process* using the *rdfs:subClassOf* relation and its semantic indicates the type of process (creation of a product of type A).
  This concept has a domain specific attribute *pa:department* which represents the identifier of the department that executed the process.
- *pa:Assemble*: this concept extends the concept *ebtic-bpm:Task* (with the *rdfs:subClassOf* relation) and represents the activity of assembling a set of components (indicated by the relation *pa:useComponent*) into a new component (represented by the relation *pa:createComponent*).
- *pa:Component*: this concept represents the component that is used/created by the assembling task. This concept has an attribute *pa:serialNumber* which indicates the serial number of the component.
- *pa:Test*: this concept extends the concept *ebtic-bpm:Task* and represents the activity of testing the final component (indicated by the *pa:testedComponent* relation). As represented by the *pa:executedBy* relation, this activity is carried out by an employee.
- *pa:Employee*: this concept represents the employee which performs the test of the final product. This concept has an attribute *pa:EIN* which represents the identification number of the employee.

Table 2. The list of triples that extends the EBTIC-BPM vocabulary with domain specific information

s	p	o
pa:CreateProductA	rdfs:subClassOf	ebtic-bpm:Process
pa:Assemble	rdfs:subClassOf	ebtic-bpm:Task
pa:Test	rdfs:subClassOf	ebtic-bpm:Task
pa:Assemble	rdfs:domain	pa:useComponent
pa:useComponent	rdfs:range	pa:Component
pa:Assemble	rdfs:domain	pa:createComponent
pa:createComponent	rdfs:range	pa:Component
pa:Test	rdfs:domain	pa:testedComponent
pa:testedComponent	rdfs:range	pa:Component
pa:Test	rdfs:domain	pa:executedBy
pa:executedBy	rdfs:range	pa:Employee
pa:CreateProductA	rdfs:domain	pa:department
pa:department	rdfs:range	xs:String
pa:Component	rdfs:domain	pa:serialNumber
pa:serialNumber	rdfs:range	xs:Integer
pa:Employee	rdfs:domain	pa:EIN
pa:EIN	rdfs:range	xs:Integer

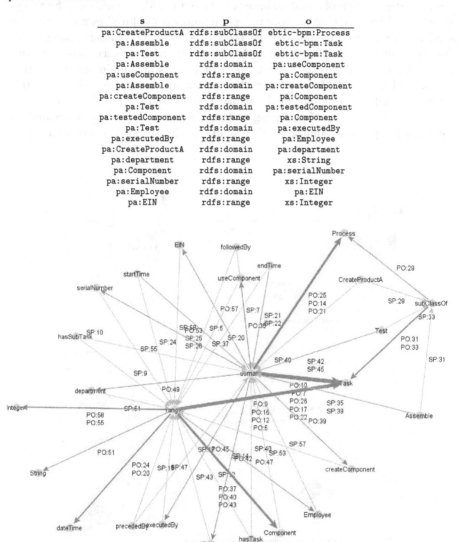

Fig. 4. The RDF graph resulting by adding to the existing graph the the triples in Table 2

This domain specific information is used to extend the basic process model by appending the triples in Table 2 to the ones present in the triple store (the triples in Table 1 that will be present after the bootstrap procedure). Figure 4 shows the complete RDF graph resulting by the union of the triples in Table 1 and in Table 2.

3.2   Architecture of a Sample Deployment

A typical deployment of a system based on the data model described so far is represented in Figure 5. One or more process monitor is deployed as a non invasive application and used to monitor the process activity; non invasive means that the deployment of the process monitor will require none or very limited change in the existing system. The process monitor creates and maintains the extension of the EBTIC-BPM vocabulary while capturing process execution data: this extension need to be present in the triple store as well as the EBTIC-BPM vocabulary: it can be provided during system bootstrap procedure or created on the fly, by the process monitor itself by analysing the process execution data, and

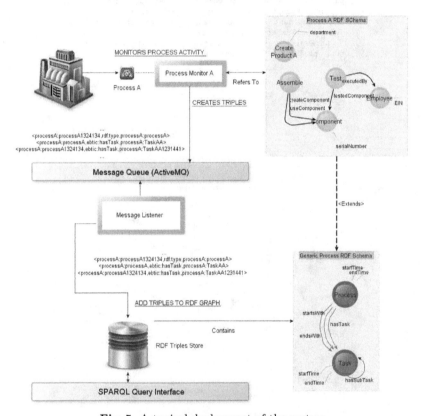

Fig. 5. A typical deployment of the system

Table 3. An example of a log file contining process activity

TimeStamp	TransactionBegin	ID	TN	Dep.	SN	EIN
2012-03-14T03:55:46	2012-03-14T03:18:56	01	Assemble	DBX	23442	
2012-03-14T04:15:06	2012-03-14T04:01:56	01	Assemble	DBX	21232	
2012-03-14T04:55:26	2012-03-14T04:18:56	01	Assemble	DBX	21232	
2012-03-14T05:15:16	2012-03-14T04:56:58	01	Test	DBX	21232	760506653
2012-03-14T06:55:26	2012-03-14T06:18:56	01	Assemble	DBX	21232	
2012-03-14T06:58:25	2012-03-14T07:15:26	01	Test	DBX	21232	760506653

Table 4. The triples defining one instance of execution of a process, as created by a monitoring application

s	p	o
pa:01	rdf:type	pa:CreateProductA
pa:01	pa:department	DBX"^^xsd:string
pa:01	ebtic-bpm:startTime	2012-03-14T03:18:56+04:00"^^xsd:date
pa:0045	rdf:type	pa:Assemble
pa:0045	ebtic-bpm:startTime	2012-03-14T03:18:56+04:00"^^xsd:date
pa:0098	rdf:type	pa:Component
pa:0098	pa:serialNumber	23442"^^xsd:integer
pa:0045	pa:createComponent	pa:0098
pa:0045	ebtic-bpm:endTime	2012-03-14T03:55:46+04:00"^^xsd:date
pa:01	ebtic-bpm:hasTask	pa:0045
pa:0045	ebtic-bpm:followedBy	pa:0046
pa:0046	ebtic-bpm:followedBy	pa:0047
pa:0047	ebtic-bpm:precededBy	pa:0046
pa:0056	ebtic-bpm:precededBy	pa:0049
pa:0046	rdf:type	pa:Assemble
pa:0046	ebtic-bpm:startTime	2012-03-14T04:01:56+04:00"^^xsd:date
pa:0099	rdf:type	pa:Component
pa:0099	pa:serialNumber	21232"^^xsd:integer
pa:0046	pa:useComponent	pa:0098
pa:0046	pa:createComponent	pa:0099
pa:0046	ebtic-bpm:endTime	2012-03-14T04:15:06+04:00"^^xsd:date
pa:0047	rdf:type	pa:Assemble
pa:01	ebtic-bpm:hasTask	pa:0046
pa:0047	ebtic-bpm:startTime	2012-03-14T04:18:56+04:00"^^xsd:date
pa:0047	pa:useComponent	pa:0099
pa:0048	ebtic-bpm:precededBy	pa:0047
pa:0047	ebtic-bpm:followedBy	pa:0048
pa:0048	ebtic-bpm:followedBy	pa:0049
pa:0049	ebtic-bpm:followedBy	pa:0056
pa:0046	ebtic-bpm:precededBy	pa:0045
pa:0049	ebtic-bpm:precededBy	pa:0048
pa:0047	pa:createComponent	pa:0099
pa:0047	ebtic-bpm:endTime	2012-03-14T04:55:26+04:00"^^xsd:date
pa:0048	rdf:type	pa:Test
pa:01	ebtic-bpm:hasTask	pa:0049
pa:0048	ebtic-bpm:startTime	2012-03-14T04:56:58+04:00"^^xsd:date
pa:0048	pa:testedComponent	pa:0099
pa:00129	rdf:type	pa:Employee
pa:00129	pa:EIN	760506653"^^xsd:integer
pa:0048	pa:executedBy	pa:00129
pa:0048	ebtic-bpm:endTime	2012-03-14T05:15:16+04:00"^^xsd:date
pa:0049	rdf:type	pa:Assemble
pa:01	ebtic-bpm:hasTask	pa:0047
pa:0049	ebtic-bpm:startTime	2012-03-14T06:18:56+04:00"^^xsd:date
pa:0049	pa:useComponent	pa:0099
pa:0049	pa:createComponent	pa:0099
pa:0049	ebtic-bpm:endTime	2012-03-14T06:55:26+04:00"^^xsd:date
pa:0056	rdf:type	pa:Test
pa:01	ebtic-bpm:hasTask	pa:0048
pa:0056	ebtic-bpm:startTime	2012-03-14T06:58:25+04:00"^^xsd:date
pa:0056	pa:testedComponent	pa:0099
pa:0056	pa:executedBy	pa:00129
pa:01	ebtic-bpm:hasTask	pa:0056
pa:0056	ebtic-bpm:endTime	2012-03-14T07:15:26+04:00"^^xsd:date
pa:01	ebtic-bpm:endTime	2012-03-14T07:15:26+04:00"^^xsd:date

submitted gradually to the triple store. When the process monitor intercept process activity information, it creates the triples representing such activity and continuously sends the triples to a message queue (for example Apache ActiveMQ[5]).

---

[5] http://activemq.apache.org/

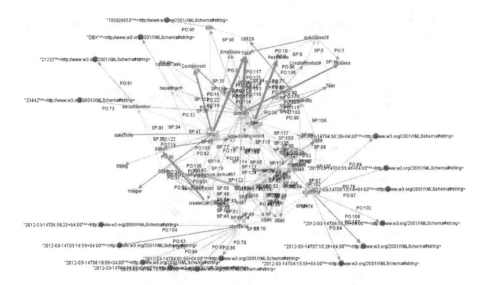

Fig. 6. The RDF graph resulting by the union of the triples of Table 1, 2 and 4

A message listener continuously receives the triples sent to the message queue by the various process monitors and inserts the triples into the triple store. The triple store exposes a SPARQL query service allowing external applications to query the continuously updated process execution graph.

As discussed, the monitoring application is responsible of creating the extensions of the EBTIC-BPM vocabulary and once a process instance is executing, the process monitor captures this information and creates the corresponding triples to be sent to the triple store.

For the sake of simplicity we will not consider in this paper named graphs or contexts of a triple. We assume the triples created by the process monitors will contribute to compose the same RDF graph. We will demonstrate that even without the use of contexts we can distinguish between different process models in the same graph. Introducing named graphs or contexts will add separation between RDF graphs, but the approach described is still valid.

In this deployment, the process monitoring application generates a flow of triples representing the process activity. The logic used to translate process activity into triples is contained in the Process Monitor and it varies depending on the process. Imagine a log file continuously updated by a workflow management system containing process activity as the on in Table 3.

Every line in the log represent a process activity; every time a line is added to the log, the process monitor will analyse the line and crate the respective triples. The final list of triples representing the log execution of Table 3 is reported in Table 4.

Figure 6 shows the complete RDF graph containing all the nodes and connections representing the execution of a process; as it is possible to notice, already with only one process execution the graph starts to become complex.

Once the triples are present in the triple store, it will be possible for other applications such as business process mining or monitoring applications to extract and analyse business process information through the SPARQL interface.

The EBTIC-BPM vocabulary introduced in section 3 provides a generic business process domain vocabulary that allows to define SPARQL queries to be used for data discovery process, but it is necessary to ensure that the concepts, relations and attributes in Figure 1 are always present in the graph when the process monitors become operative. Therefore we assume that the system will be initialised by a bootstrap procedure that will insert the triples of Table 1 in the triple store.

4   A SPARQL-Based Discovery Process

As a proof of concept application we developed a business process visualisation tool, which is based on a set of predefined SPARQL queries defined with the assumption that the EBTIC-BPM vocabulary is present in the graph. Hence the queries are independent with respect to the domain specific information of the process to analyse.

This application connects to a SPARQL endpoint and is used to display business process flows under different point of views. It has been developed from a set of recurrent requirements gathered with the experience our team developed from down-streaming of the business process mining tool [3]. These requirements were focused on a more flexible visualisation of the process workflow and the possibility to monitor real time process executions. The use of the data model described in this article allows to fulfil both requirements successfully.

This tool will initially display the list of domain specific processes that are available in the triple store, which the user can select and obtain the instances of that process, inspect their attributes, and list of tasks with their attributes, related concepts, and subtasks. Also by clicking on a process instance the application will display the workflow diagram of the selected instance of execution. All the interactions of the user with the application are translated into SPARQL queries that are submitted to the triple store. This application processes the results and presents them to the user according to the selected action.

These SPARQL queries are defined only with knowledge of EBTIC-BPM vocabulary, but we will now demonstrate that the tool can visualise and make use of domain specific information without prior knowledge of it.

Once the application is started, a set of SPARQL queries is executed in order to obtain some information about the processes that are present in the store; the following query:

SELECT ?process   (1)
WHERE { ?process rdfs:subClassOf ebtic-bpm:Process.}

returns the concepts that extend the *ebtic-bpm:Process* concept. Each of these concepts represents a different process type that is present in the RDF graph, in case of the current example the results in the variable ?process will contain

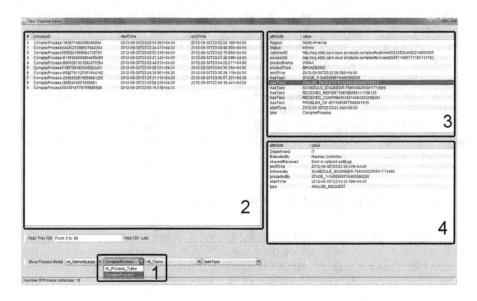

Fig. 7. The main interface of the business process visualisation tool

pa:CreateProductA. The results of this query will be displayed in a list (area 1 in Figure 7) and the user will be able to select which type of processes to analyse. Once the user selects an entry in the process list, a second query will be executed:

SELECT ?processID ?startTime ?endTime   (2)

WHERE { ?processID rdf:type pa:CreateProductA.

      ?processID ebtic-bpm:startTime ?startTime.

      OPTIONAL { ?processID ebtic-bpm:endTime ?endTime.}}

this query returns information (?processID ?startTime ?endTime) about the instances of the process *pa:CreateProductA*, in case the user selected such process from the list populated with the results of the previous query. In case the user did not select any specific process type then the query that will be submitted will be slightly different:

SELECT ?processID ?startTime ?endTime   (3)

WHERE { ?processID rdf:type ?process.

      ?process rdfs:subClassOf ebtic-bpm:Process.

      ?processID ebtic-bpm:startTime ?startTime.

      OPTIONAL { ?processID ebtic-bpm:endTime ?endTime.}}

this modified version combines the query 1, used to retrieve the concepts extending the basic process concept and the query 2 which is used to retrieve the instances of a specific process concept, thus the query will return information about all the process instances that are present in the graph; the query 3 can be slightly simplified in case the triple store supports RDFS inference.

The results of this second query will be used to populate a table in the graphical interface of the application (area 2 in Figure 7), where the user will be able to see the list of instances of the selected process (pa:CreateProductA in this case). The OPTIONAL keyword indicates that the ending time of a process may not be present, in case the process is still executing (as it is possible to notice in the last row of the table in area 2 in Figure 7). In case OPTIONAL is not present the results will contain only process instances that have terminated.

At this point the user, by interacting with this list can select one process instance, as an example pa:01; this action will fire another query:

$$\text{SELECT ?attribute ?value} \qquad (4)$$
$$\text{WHERE \{ pa:01 ?attribute ?value.}$$
$$\text{FILTER (?attribute != rdf:type)\}}$$

which will populate another table with the names and values of all the attributes of the selected process and the list of its tasks (area 3 in Figure 7).

It is important to stress on the fact that also in this case the query will return all the process attributes and tasks of that specific process execution, without any knowledge of the process meta-data. The FILTER keyword is used to remove the attributes of type rdf:type from the results if they are not meaningful for the final user. The use of pa:01 in the first triple pattern is used to ensure that the results returned are relevant to the selected process instance.

The user can now select a row in the resulting table, this action will execute the query 4 modified by replacing the pa:01 element with the value of the row selected by the user. In case the value represents a task instance, the results of this query will populate a table with the names and the values of all the attributes of the selected task (area 4 in Figure 7)

A more interesting behaviour occurs when the user double clicks on one process instance to obtain the process workflow. This action will be converted by the application into the following SPARQL query:

$$\text{SELECT ?PID ?startTime ?endTime ?taskID} \qquad (5)$$
$$\text{?taskType ?follBy ?precBy}$$
$$\text{WHERE \{ ?PID ebtic-bpm:hasTask ?taskID.}$$
$$\text{?taskID rdf:type ?taskType.}$$
$$\text{?PID ebtic-bpm:startTime ?startTime.}$$
$$\text{OPTIONAL \{?PID ebtic-bpm:endTime ?endTime.\}.}$$
$$\text{OPTIONAL \{ ?taskID ebtic-bpm:followedBy ?follBy.\}.}$$
$$\text{OPTIONAL \{ ?taskID ebtic-bpm:precededBy ?precBy.\}.}$$
$$\text{FILTER (?PID = pa:01)\}}$$

where pa:01 is the selected process instance.

The output of query 5, in case the RDF graph contains the triples of Table 1, 2 and 4 is reported in Table 5

Table 5. A sample result set from the execution of query 5

PID	startTime	endTime	taskID	taskType	follBy	precBy
pa:01	2012-03-14T03:18:56+04:00	2012-03-14T07:15:26+04:00	pa:0045	pa:Assemble	pa:0046	
pa:01	2012-03-14T03:18:56+04:00	2012-03-14T07:15:26+04:00	pa:0046	pa:Assemble	pa:0047	pa:0045
pa:01	2012-03-14T03:18:56+04:00	2012-03-14T07:15:26+04:00	pa:0047	pa:Assemble	pa:0048	pa:0046
pa:01	2012-03-14T03:18:56+04:00	2012-03-14T07:15:26+04:00	pa:0048	pa:Test	pa:0049	pa:0047
pa:01	2012-03-14T03:18:56+04:00	2012-03-14T07:15:26+04:00	pa:0049	pa:Assemble	pa:0056	pa:0048
pa:01	2012-03-14T03:18:56+04:00	2012-03-14T07:15:26+04:00	pa:0056	pa:Test		pa:0049

The result of this query is processed by a method that returns a graphical representation of the process built using the information in the ?followedBy and ?precededBy variables. The method generates a standard GraphML [17] document that can be consumed by any graphical library supporting such standard (Jung[6], yFiles[7], jsPlumb[8], D3[9] just to cite few of them). The output will be returned to the user in a new window as in the example of Figure 8. Our tool makes use of the yFiles library to display the GraphML document. All the workflow diagrams presented in this paper has been generated by our tool; in the figures the window frame has been cropped out.

The process workflow in Figure 8 is an interactive object that the user can manipulate: as an example switching the task-based view by selecting alternative ways to present the process. This can be done at *process* or *task* level.

4.1   Process Level View Interaction

This aspect refers to the interaction of the user with the graphical representation of the global process workflow. The right click of the user on any point of the window (which is not a task), is translated in a query that is used to retrieve all the attributes and objects related to all the tasks present in the workflow. This is an important feature of this approach because it ensures that the user will only be able to choose between task attributes that are present in the process instance the user is interacting with. This is encoded in the following query:

$$\text{SELECT DISTINCT ?attribute} \tag{6}$$

```
WHERE { pa:01 ebtic-bpm:hasTask ?taskID.
?taskID rdf:type ?Task.
?Task rdfs:subClassOf ebtic-bpm:Task.
?Task rdfs:domain ?attribute.}
```

The results of this query will populate a pop-up box where the user can select an element that will be used to switch the process workflow from a task-based representation to an attribute-based representation, according to the attribute selected. In case of process level view interaction, if the selected attribute is not present in a task the usual task-based representation will be used for that task.

---

[6] http://jung.sourceforge.net/
[7] http://www.yworks.com/en/products_yfiles_about.html
[8] http://jsplumb.org
[9] http://d3js.org/

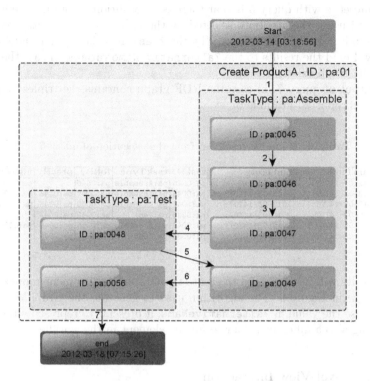

Fig. 8. The process workflow of a process instance

This behaviour is automatically encoded in the query that will be fired once the user selects an attribute from the pop-up list. The query is similar to the previous one, with one additional variable `?alternativeName` that will be used to replace the task name in case the attribute used to switch the workflow is present in the task. For example, if the user selects the element `pa:createComponent` from the pop up list; the SPARQL query that will be execute will be:

```
SELECT ?PID ?startTime ?endTime ?taskID                        (7)
       ?taskType ?follBy ?precBy ?altName
WHERE { ?PID ebtic-bpm:hasTask ?taskID.
        ?taskID rdf:type ?taskType.
        ?PID ebtic-bpm:startTime ?startTime.
OPTIONAL {?PID ebtic-bpm:endTime ?endTime.}.
OPTIONAL { ?taskID ebtic-bpm:followedBy ?follBy.}.
OPTIONAL { ?taskID ebtic-bpm:precededBy ?precBy.}.
OPTIONAL { ?taskID pa:createComponent ?altName.}.
FILTER (?PID = pa:01)}
```

The difference with query 5 is that there is an additional optional triple pattern created using the attribute selected by the user (pa:createComponent in this case) linked to the task instance (by the ?taskID variable). This additional pattern will bind the results to the ?alternativeName variable in case the path in the RDF graph exists.

The output of query 7, in case the RDF graph contains the triples of Table 1, 2 and 4 is reported in Table 6.

Table 6. A sample result set from the execution of query 5

PID	startTime	endTime	taskID	taskType	follBy	precBy	altName
pa:01	2012-03-14T07:...	2012-03-14T07:...	pa:0045	pa:Assemble	pa:0046		pa:0098
pa:01	2012-03-14T07:...	2012-03-14T07:...	pa:0046	pa:Assemble	pa:0047	pa:0045	pa:0099
pa:01	2012-03-14T07:...	2012-03-14T07:...	pa:0047	pa:Assemble	pa:0048	pa:0046	pa:0099
pa:01	2012-03-14T07:...	2012-03-14T07:...	pa:0048	pa:Test	pa:0049	pa:0047	
pa:01	2012-03-14T07:...	2012-03-14T07:...	pa:0049	pa:Assemble	pa:0056	pa:0048	pa:0099
pa:01	2012-03-14T07:...	2012-03-14T07:...	pa:0056	pa:Test		pa:0049	

The results will be processed to create the GraphML document representing the workflow in Figure 9. The algorithm used to generate the GraphML document will use the value in ?alternativeName, if present, to represents the task name, using also a different colour to fill the element in the workflow.

4.2   Task Level View Interaction

This second aspect refers to the recursive interaction of the user with single task type attributes in order to modify the graphical representation of the process workflow.

Imagine a situation for which the user is interested in a certain attribute of a specific task type and another attribute of another task type.

The user, by right clicking on a task, will fire a query very similar to query 6 which will be used to retrieve all the attributes and objects related to that specific task. This is done by replacing the variable ?Task in the query with the specific task (pa:createComponent), consequently narrowing the set of results.

These results are used to populate a pop-up box as, described above, and the user can select an attribute that will be used to replace the traditional task-based representation for that specific task.

The action of clicking on an attribute will be translated to a query conceptually similar to query 7, with the difference that this time we need to make wide use of the UNION construct in order to compose the results of different queries: one query that will be focused on the alternative value referred to the task selected by the user, while a second query that will take care of the tasks that are

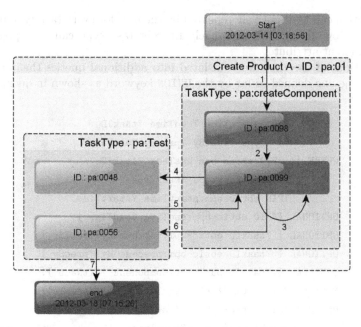

Fig. 9. The process workflow using switched view by the id of the component created by the task

not of the type selected by the user, which will be displayed with the traditional task-based view. This behaviour is defined by the following query:

```
SELECT ?PID ?startTime ?endTime ?taskID                    (8)
        ?taskType ?follBy ?precBy ?altName
WHERE {{ ?PID ebtic-bpm:hasTask ?taskID.
          ?taskID rdf:type ?taskType.
          ?PID ebtic-bpm:startTime ?startTime.
OPTIONAL {?PID ebtic-bpm:endTime ?endTime.}.
OPTIONAL { ?taskID ebtic-bpm:followedBy ?follBy.}.
OPTIONAL { ?taskID ebtic-bpm:precededBy ?precBy.}.
OPTIONAL { ?taskID pa:createComponent ?altName.}.
FILTER (?PID = pa:01 && ?taskType = pa:Assemble)}
UNION { ?PID ebtic-bpm:hasTask ?taskID.
          ?taskID rdf:type ?taskType.
          ?PID ebtic-bpm:startTime ?startTime.
OPTIONAL {?PID ebtic-bpm:endTime ?endTime.}.
OPTIONAL { ?taskID ebtic-bpm:followedBy ?follBy.}.
OPTIONAL { ?taskID ebtic-bpm:precededBy ?precBy.}.
FILTER (?PID = pa:01 && ?taskType != pa:Assemble)}}
```

This action can be iteratively repeated by the user for every task type present in the workflow diagram, so that each different task type can be represented using a different attribute.

The actions of the user are translated into additional queries that are connected to the one defined so far by the UNION keyword as shown in query 9.

```
SELECT ?PID ?startTime ?endTime ?taskID                          (9)
          ?taskType ?follBy ?precBy ?altName
WHERE {{ ?PID ebtic-bpm:hasTask ?taskID.
          ?taskID rdf:type ?taskType.
          ?PID ebtic-bpm:startTime ?startTime.
OPTIONAL {?PID ebtic-bpm:endTime ?endTime.}.
OPTIONAL { ?taskID ebtic-bpm:followedBy ?follBy.}.
OPTIONAL { ?taskID ebtic-bpm:precededBy ?precBy.}.
OPTIONAL { ?taskID pa:createComponent ?altName.}.
FILTER (?PID = pa:01 && ?taskType = pa:Assemble)}
UNION { ?PID ebtic-bpm:hasTask ?taskID.
          ?taskID rdf:type ?taskType.
          ?PID ebtic-bpm:startTime ?startTime.
OPTIONAL {?PID ebtic-bpm:endTime ?endTime.}.
OPTIONAL { ?taskID ebtic-bpm:followedBy ?follBy.}.
OPTIONAL { ?taskID ebtic-bpm:precededBy ?precBy.}.
OPTIONAL { ?taskID pa:executedBy ?altName.}.
FILTER (?PID = pa:01 && ?taskType = pa:Test)}
UNION { ?PID ebtic-bpm:hasTask ?taskID.
          ?taskID rdf:type ?taskType.
          ?PID ebtic-bpm:startTime ?startTime.
OPTIONAL {?PID ebtic-bpm:endTime ?endTime.}.
OPTIONAL { ?taskID ebtic-bpm:followedBy ?follBy.}.
OPTIONAL { ?taskID ebtic-bpm:precededBy ?precBy.}.
FILTER (?PID = pa:01 && ?taskType != pa:Assemble
          && ?taskType != pa:Test)}}
```

The output of query 9, in case the RDF graph contains the triples of Table 1, 2 and 4 is reported in Table 7.

The results will be processed to create the GraphML document representing the workflow in Figure 10.

This set of queries demonstrates how the visualiser application is able to extract and make use of domain specific information to create useful user interfaces and rich process workflows, by automatically creating SPARQL queries with only the initial knowledge of the EBTIC-BPM vocabulary. As previously pointed out

Table 7. A sample result set from the execution of query 6.

PID	startTime	endTime	taskID	taskType	follBy	precBy	altName
pa:01	2012-03-14T03:...	2012-03-14T07:...	pa:0045	pa:Assemble	pa:0046		pa:0098
pa:01	2012-03-14T03:...	2012-03-14T07:...	pa:0046	pa:Assemble	pa:0047	pa:0045	pa:0099
pa:01	2012-03-14T03:...	2012-03-14T07:...	pa:0047	pa:Assemble	pa:0048	pa:0046	pa:0099
pa:01	2012-03-14T03:...	2012-03-14T07:...	pa:0049	pa:Assemble	pa:0056	pa:0048	pa:0099
pa:01	2012-03-14T03:...	2012-03-14T07:...	pa:0048	pa:Test	pa:0049	pa:0047	pa:00129
pa:01	2012-03-14T03:...	2012-03-14T07:...	pa:0056	pa:Test		pa:0049	pa:00129

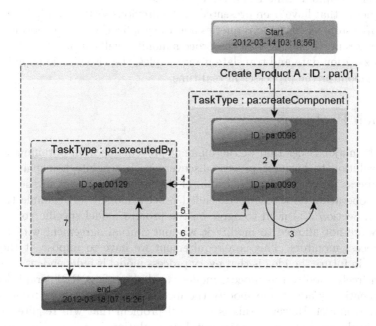

Fig. 10. The process workflow using switched view by used component

this vocabulary is the basic set of information shared between applications that monitor business process execution and submit the triples to the triple store, and the applications that make use of such information.

5   Real Time Aspects

Another important aspect of the business process model presented in this article is the possibility to continuously add information to the graph. This is an important feature of RDF representation, enabled by the fact that every triple is a valid RDF piece of information that identify nodes and connections in the RDf graph. Another important feature of RDF is that both schema and instance-level information is stored in the same graph and that a SPARQL query can return any point in the graph.

Answering queries in SPARQL can be simply translated into identifying all the paths in the graph that satisfy the query pattern. Hence, continuously adding nodes and connection to the graph generates new paths in the graph that can match queries previously executed.

EBTIC[10] developed a triple store that allows client applications to register queries, so that the client is notified with a new result whenever there is data inserted in the system closing a path that satisfies the registered query. In case the triple store does not support such a feature, a timer can be defined and the queries to monitor can be executed continuously providing the difference between the results at time t and t-1 to the client.

The queries that have been presented in the previous section can be registered in the triple store as continuous queries and the application will be notified with every new result. Assuming that the process monitor will continuously intercept process execution data and translate it into triples, the visualisation application is able to monitor the processes in real-time.

6   Open Problems

One of the major issues that at the moment prevents us from obtaining a fully flexible solution to business process discovery with absolute transparency between process monitors and the graph, is the assumption that all the process monitors operating on the same extended process model (for example the one defined in Section 3.1) need to agree on the process model specific vocabulary and they are not allowed to modify it without mutual agreement with all the other process monitors. This is a requirement we have to impose at the moment, but in future the idea is that every process model is independent and that they can freely modify the process model vocabulary. The RDF graph ideally will automatically adapt and modify the incoming triples in order to adapt to the evolved model. However this is an open problem that will require further research in the field of collaborative ontology evolution.

7   Conclusions

In this article we presented an extremely extensible and flexible data representation model oriented towards real time business process monitoring and discovering; the model is based on Resource Description Framework (RDF) standard that allows independence between applications that generate business process data and applications that consume it. The process is represented as a labelled oriented graph defined by a set of triples as defined in Section 3.

Applications that monitor business processes will extend this basic set of information with their domain specific one at run-time by entering the corresponding triples in the triple store.

We finally demonstrated that this approach allows process discovery and analysis of domain specific extensions that may also be created at run time by third party applications just with the use of SPARQL queries. Future work on this

---

[10] http://www.ebtic.org

direction will be to develop a set of non-invasive monitoring and analytical applications that will allow us to deploy and test this approach within an enterprise-scale environment.

References

1. van der Aalst, W.M.P., Weijters, T., Maruster, L.: Workflow mining: Discovering process models from event logs. IEEE Transactions on Knowledge and Data Engineering 16(9), 1128–1142 (2004)
2. van der Aalst, W.M.P., van Dongen, B.F., Herbst, J., Maruster, L., Schimm, G., Weijters, A.J.M.M.: Workflow mining: a survey of issues and approaches. Data Knowl. Eng. 47, 237–267 (2003)
3. Taylor, P., Leida, M., Majeed, B.: Case study in process mining in a multinational enterprise. In: Aberer, K., Damiani, E., Dillon, T. (eds.) SIMPDA 2011. LNBIP, vol. 116, pp. 134–153. Springer, Heidelberg (2012)
4. van der Aalst, W.M.P., et al.: Process mining manifesto. In: Daniel, F., Barkaoui, K., Dustdar, S. (eds.) BPM Workshops 2011, Part I. LNBIP, vol. 99, pp. 169–194. Springer, Heidelberg (2012)
5. Hayes, P., McBride, B.: Resource description framework (rdf). Recommendation, W3C (2004)
6. Prud'hommeaux, E., Seaborne, A.: Sparql query language for rdf. Recommendation, W3C (2008)
7. Fischer, M.: Aris process performance manager. In: Bause, F., Buchholz, P. (eds.) MMB, pp. 307–310. VDE Verlag (2008)
8. Weber, I., Hoffmann, J., Mendling, J.: Semantic business process validation. Technical report (2008)
9. Roman, D., Keller, U., Lausen, H., de Bruijn, J., Lara, R., Stollberg, M., Polleres, A., Feier, C., Bussler, C., Fensel, D.: Wsmo - web service modeling ontology. In: DERI Working Draft 14, vol. 1, pp. 77–106. Digital Enterprise Research Institute (DERI), IOS Press, BG Amsterdam (2005)
10. Azvine, B., Cui, Z., Nauck, D.D., Majeed, B.: Real time business intelligence for the adaptive enterprise. In: 8th IEEE International Conference on E-Commerce Technology - 3rd IEEE International Conference on Enterprise Computing, E-Commerce, and E-Services, CEC-EEE 2006. IEEE Computer Society, Washington, DC (2006)
11. Brickley, D., Guha, R., McBride, B.: Rdf vocabulary description language 1.0: Rdf schema. Recommendation, W3C (2004)
12. Carroll, J.J., Bizer, C., Hayes, P., Stickler, P.: Named graphs. Web Semant. 3(4), 247–267 (2005)
13. Rohloff, K., Dean, M., Emmons, I., Ryder, D., Sumner, J.: An evaluation of triplestore technologies for large data stores. In: Meersman, R., Tari, Z. (eds.) OTM-WS 2007, Part II. LNCS, vol. 4806, pp. 1105–1114. Springer, Heidelberg (2007)
14. Shi, H., Maly, K., Zeil, S., Zubair, M.: Comparison of ontology reasoning systems using custom rules. In: Proceedings of the International Conference on Web Intelligence, Mining and Semantics, WIMS 2011, pp. 16:1–16:9. ACM, New York (2011)
15. Bizer, C., Schultz, A.: Berlin sparql benchmark (bsbm). Benchmark, Freie Universität Berlin (2011)
16. Bizer, C., Cyganiak, R., Heath, T.: How to publish Linked Data on the Web (2007)
17. Brandes, U., Eiglsperger, M., Herman, I., Himsolt, M., Marshall, M.S.: GraphML progress report, structural layer proposal. In: Mutzel, P., Jünger, M., Leipert, S. (eds.) GD 2001. LNCS, vol. 2265, pp. 501–512. Springer, Heidelberg (2002)

Combination of Process Mining and Simulation Techniques for Business Process Redesign: A Methodological Approach

Santiago Aguirre, Carlos Parra, and Jorge Alvarado

Industrial Engineering Department, Pontificia Universidad Javeriana,
Bogotá, Colombia
{saguirre,carlos.parra,jorge.alvarado}@javeriana.edu.co

Abstract. Organizations of all sizes are currently supporting their performance on information systems that record the real execution of their business processes in event logs. Process mining tools analyze the log to provide insight on the real problems of the process, as part of the diagnostic phase. Nonetheless, to complete the lifecycle of a process, the latter has to be redesigned, a task for which simulation techniques can be used in combination with process mining, in order to evaluate different improvement alternatives before they are put in practice. In this context, the current work presents a methodological approach to the integration of process mining and simulation techniques in a process redesign project.

Keywords: process mining, data mining, simulation, process redesign, BPM.

1 Introduction

Information systems have become the backbone of most organizations. Without them, companies could not sell products or services, purchase materials, pay suppliers or submit their tax reports. These systems record valuable information about process execution on event logs containing activities, originators, timestamps and case data. This information can be extracted and analyzed to produce useful knowledge for organizations to diagnose and improve their business processes. This is called process mining [1] .

Process mining is a discipline that aims to discover, monitor and improve business processes by extracting knowledge from information systems event logs [2], making use of data mining techniques. Event logs record information about real business process execution and are available in Process Aware Information Systems (PAIS) such as BPM, ERP, CRM, Workflow Management Systems, etc. [3]. Process mining is, therefore, a recent discipline that lies between data mining and process modeling and analysis [1].

The ultimate goal of process mining is to generate useful knowledge for organizations to understand and improve their business processes mainly through the application of data-mining-based tools. Figure 1 shows the three components of process mining [2]: Discovery, Conformance and Enhancement.

P. Cudre-Mauroux, P. Ceravolo, and D. Gašević (Eds.): SIMPDA 2012, LNBIP 162, pp. 24–43, 2013.
© International Federation for Information Processing 2013

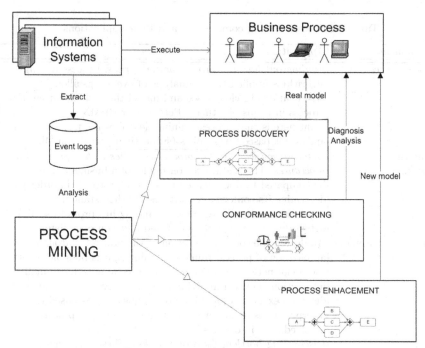

Fig. 1. Process mining components

Table 1 details the organizational enhancement possibilities offered by process mining through its three components.

For its part, Business Process Simulation provides techniques for testing solutions before their actual implementation. Both simulation and process mining contribute to Business Process Lifecycle [9] (Figure 2), which starts with business process design based on customer and stakeholder requirements. Next, the implementation stage comprises business rules and policy definition as well as computer platform configuration. Then, the process enters its actual execution stage. Later, in the monitoring and analysis stage, the process is optimized by measuring and analyzing its performance indicators. Table 2 details the contribution of simulation and process mining to each stage of the cycle.

According to the authors of the Process Mining Manifesto[1], one of the challenges that must be addressed to improve the usability of process mining is its integration with other methodologies and analysis techniques. A clear example is provided by simulation tools, which are likely to complement process diagnosis and analysis by testing alternative process mining implementation scenarios as part of the business process lifecycle.

Most simulation techniques have been applied to production and logistics, where process routes are predefined and can therefore be more easily modeled. However, in service processes such as complaint appraisal and response, there

Table 1. Process Mining components and their applications

Component	Application
Process Discovery	*Finding out how the process actually runs.* Process mining algorithms applied to the analysis of event logs allow organizations to clearly see and model the real execution of a process in terms of either a Petri net or BPMN notation. The point here is that process mining describes the real situation and is not based on people's (subjective) perception [4].
Conformance Checking	*Determining whether the process complies with regulations and procedures.* The real execution model of a business process can be compared to documented procedure protocols in order to determine its conformance with established standards, regulations and policies. Process mining has proved useful for detecting potential sources of fraud and non-compliance [5].
Process Enhacement	*Analyzing the social interaction of the process.* Through the application of process mining techniques, it is possible to assess the social network supporting the process, in order to analyze interactions between individuals and discover loops that may delay its execution [6]. These techniques are also used to interpret roles in the process as an example group of users involved only in one task. *Discovering bottlenecks (bottlenecks).* These techniques allow finding actual bottlenecks on which action can be taken to improve process implementation. *Predicting specific time cycles.* Certain data mining techniques such as decision trees facilitate the prediction of the remaining execution time of a running process [7], [8].

can be many variations or process routes depending on the type of complaint. For this reason, it is important to start by analyzing the information system's event log in order to reach a realistic model, rather than an idealized version of the process. The necessary parameters to build such a model can be supplied by process mining.

Furthermore, a series of methodological approaches have been developed for business process redesign and for the application of simulation and process mining. BP trends [10] proposes five general stages for a process redesign effort: 1) Project Understanding, 2) Business Process Analysis, 3) Business Process Redesign, 4) Business Process Redesign Implementation and 5) Redesigned Business Process Roll Out. Although this methodology constitutes a valuable approach, it needs to be complemented with specific tools such as simulation and process mining.

The current paper presents a methodological approach to process redesign, based on a combination of simulation techniques and both data and process mining tools, together with those of the understanding phase of the BP trends

Table 2. Contributions of simulation and process mining to Business Process Lifecycle

Phase	Contribution of process mining	Contribution of simulation
Re(Design)	The real process model, which is discovered by process mining techniques, is an important input for process design or redesign.	Through simulation it is possible to perform "what if" analyses of the different process design or redesign options.
Implementation and Execution	In the implementation phase, process mining is used to verify that the process complies with business policies and rules. It is also possible to predict the remaining execution time of a running case.	Having been tested and improved through simulation, business processes are implemented in this phase.
Monitoring and analysis	In the analysis phase, process mining is used for identifying loops and bottlenecks, and for further checking for conformance with business rules.	Simulation allows business process analysis to eliminate bottlenecks and improve throughput times.

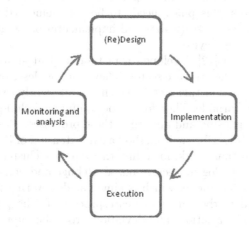

Fig. 2. Business Process Lifecycle

Business Process Redesign methodology [10]. Section 2 contains a complete review of the state of the art and related works. Section 3 provides a detailed explanation of the methodological approach and Section 4 describes the case study to which the method was applied. Finally, Section 5 draws the conclusions and future work.

2   Literature Review and Related Works

The literature review focuses on previous methodological developments intended
not only for the application of process mining to business process improvement,
but for the combination of process mining and simulation as well.

2.1   Methodologies for Process Mining

Bozcaya [11] proposed a methodology for applying process mining to business
process diagnosis, based on three perspectives: control flow, performance and
organizational analysis. The method in question starts with Log preparation,
which includes event log extraction, interpretation and transformation, in order
to determine the activities and their sequence. The next step is to inspect and
clean the event log data to eliminate cases with missing data. Once the log has
been cleaned, the control flow analysis is performed for conformance checking
against procedures through the application of discovery techniques like alpha,
fuzzy of genetic algorithms.

As a next step, these same authors propose performance analysis in order
to discover business process bottlenecks and delays. Finally, the social network
algorithms are used to apply an organizational analysis aimed not only at determining role interactions involved in process execution, but at discovering loops
that might be delaying process cycling time. This method was applied to a case
study and constitutes an important step ahead in the diagnostic phase of process
redesign. Nevertheless, this phase needs to be complemented with the understanding (planning), redesign (to-be) and implementation stages of a complete
business process redesign cycle.

Rebuge and Ferreira [12] developed a methodological approach to business
process analysis in the health care sector. They start by describing the complexity of business processes in this sector, which are inherently dynamic, multidis-
ciplinary and highly variable. Therefore, process mining techniques are the most
suitable ones for diagnosing and analyzing these processes.

These researchers describe their method as an extension of Bozcaya's one [11],
on which they based their work, including the sequence cluster analysis applied
by this author after the log inspection phase. Rebuge and Ferreira [12] actually
focus on this cluster technique, which is aimed at discovering process flow patterns. When applied to the emergency care process of a hospital, this method
allowed identifying all variations and deviations from the internal protocols and
guidelines of the institution, thus demonstrating the usefulness of process mining for diagnosis and analysis in these cases. As to future work, they suggest
complementing the method with additional steps such as the use of heuristics
for determining the number of clusters, on the one hand, and the establishment
of measures for evaluating the quality of the results, on the other hand.

2.2   Process Mining and Simulation

The literature on this topic presents research works and case studies in which
process mining and simulation are used in combination. Rozinat [4] uses process

mining techniques to discover business processes and, based on past executions, analyzes how data attributes influence decisions on said processes. This analysis allows finding each event's probabilities and frequencies, based on which a model is constructed and represented by a Colored Petri Net (CPN), in order to simulate different resource usage optimization and throughput time reducing alternatives.

Maruster [13]proposes a process redesign methodology based on the combination of process mining and simulation techniques, and presents its application to three case studies. Mainly supported by CPN simulation, the method consists of three phases: process performance variable definition, process analysis (as-is), and process redesign (to-be). This approach constitutes an important step forward in the integration of different tools in this field. However, it focuses on CPN simulation, thus tending to underscore the understanding phase, which is seen, according to Harmon [10], as the first stage of any process redesign project.

The current work focuses on complementing the methodology proposed by Maruster [13], by emphasizing the project understanding phase, featured by process scope analysis, process redesign goal setting and performance gap analysis. According to Vander Alast [1] one of the reasons why process mining has not been widely applied is the lack of a comprehensive methodology that is capable of linking organization Key Process Indicators (KPIs) with actual analysis and redesign efforts. The methodology proposed in this paper intends to close this gap by linking business priorities to process analysis and redesign using process mining and simulation tools. Just as well, it shows how data mining tools (e.g., decision trees) can be combined with simulation in a process redesign project. Part of this method was applied to the case study described in Section 4[1] .

3   The Redesign Project: A Methodological Approach

Including process mining and simulation tools, the development of the current methodology took into consideration both BPtrends method [10] and Maruster's [3] approach. It comprises the following phases:

- *Phase I: Project Understanding.* The goal of this phase is to gain consensus over the problem to be solved, the scope of the project and the desired goals as stated in terms of the business process indicators.
- *Phase II: Project Understanding.* The goal of this phase is to gain consensus over the problem to be solved, the scope of the project and the desired goals as stated in terms of the business process indicators.
- *Phase III: Business Process Redesign (to-be).* This phase is intended to develop and simulate the corresponding business process improvement alternatives.

---

[1] Some data about this case study has been modified forprivacy reasons.

– *Phase IV: Implementation.* The goal of the implementation phase is to put in operation the amendments in question through changes in procedures, job descriptions and work assignments.

Figure 3 and Table 3 explain the activities and tools that describe the proposed methodology.

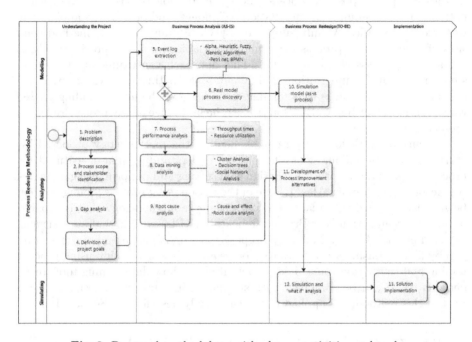

Fig. 3. Proposed methodology with phases, activities and tools

Table 3. Description of the phases and activities of the proposed methodology

Phase /Activity	Description	Tools
Phase I: Project Understanding		
1. Problem Description	The purpose of problem description is to understand and gain consensus on the reasons why the process needs to be improved (customer complains, compliance requirements, costs, throughput times).	-Process Performance Indicators.

Table 3. (*Continued*)

Phase /Activity	Description	Tools
2. Process scope and stakeholder identificatio	During this phase, the source, input, output and customer of the process are identified. The process stakeholders are established and interviewed to gain a better understanding of the performance desired for the process.	-Process scope diagram -Stakeholders diagram
3. Gap Analysis	The gap analysis identifies the actual process performance indicators (as-is) and establishes the desired process performance indicators (to –be) based on process vision, benchmarking and stakeholder expectations.	-Gap analysis -Benchmarking -Process Performance Indicators
4. Definition of Project Goals	The desired amendments of the to-be process turn to be the project goals.	
Phase II: Business Process Analysis (AS – IS)		
5. Event log extraction	The event log of the actual execution of the business process must be extracted from the information system (ERP, CRM, BPMS). Then, the event log is cleaned from missing data and transformed, so that it can be analyzed with data mining and process mining packages.	-Disco® software -PROM® software -SPSS statistical package
6. Real model process discovery	Through the application of process mining algorithms such as alpha mining [14], heuristic mining [15] or genetic mining [16], it is possible to automatically discover the actual process model from the event log using ProM or Disco software functionalities. This model can be represented in a Petri net, or through BPMN notation. The real process model allows visualizing bottlenecks, loops or lack of compliance.	-Disco® software -PROM® software Process mining algorithms (alpha, heuristics, genetic), Petri nets, BPMN.

Table 3. (*Continued*)

Phase /Activity	Description	Tools
7. Process Performance Analysis	The event log needs to be assessed through descriptive statistics in order to analyze process performance indicators such as activity times, idle times and standard deviations. Role assessment output is likely to show differences in personal productivity.	-Statistical Analysis (mean times, standard deviation). -SPSS® statistical package -Disco® software
8. Data mining Analysis	Data mining techniques are used to extract knowledge from process execution. Techniques such as decision trees are used to discover those variables that have greater incidence on process delays. Social Network is useful for analyzing role interactions between the people executing the process, with the aim of finding either functional loops or key roles within the process	-Cluster analysis, decision trees, social network analysis. -SPSS® statistical package
9. Root cause Analysis	Root cause analysis is useful to examine the causes of the main problems that have been discovered in the previous steps. This analysis is a simple way to organize and classify the list of possible causes and requires the knowledge of the people participating in the execution of the process	-Root cause diagram
Phase III: Business Process Redesign (TO-BE)		
10. Simulation Model	Based on the process discovered in phase 1, and on processing and waiting times calculated through the statistical analysis of the event log data, a simulation model is generated.	-Simulation -Simulation packages
11. Development of process improvement alternatives	Once the problems and causes are clear, the process improvement alternatives for the to-be process must be established to overcome the issues found in the as-is analysis phase (Phase II).	-Cost-benefit analysis

Table 3. (Continued)

Phase /Activity	Description	Tools
12. Simulation and "what if" analysis.	The process improvement alternatives are tested through simulation in order to decide on their actual implementation. This allows performing the "what if" analysis of different scenarios.	-Simulation -Simulation packages (Promodel, Arena, etc).
Phase IV: Implementation		
13. Solution implementation	The process improvement alternatives selected after the simulation test are then rolled out and put in practice.	Procedures, job descriptions, work assignments.

4 Case Study: Procurement Process at a Private University

The case study to which we applied the proposed methodology consists in the procurement process of a private university that handles approximately 15,000 purchase orders every year, with an estimated budget of $ US 50 million. The normal functioning of the University and its projects depends on the efficiency of the Procurement Department in obtaining the required goods and services.

The procurement process is supported by an ERP system[2] in which the following activities are executed: purchase requisition, requisition approval, purchase order, purchase order approval, goods receipt, invoice receipt and vendor payment.

4.1 Application of the Methodological Approach

The methodology presented in the current work was applied to this case study using process mining and simulation techniques. The following is the detailed step by step explanation of the process.

A. Phase I: Understanding the Project
In this phase, the problem is described, the gap analysis between as-is and to-be is performed, and the project goals are set.

Problem Description
Despite the support of an integrated system, the procurement process in question has been presenting problems and inconveniences such as long approval waiting times and overload of "manual" documents and activities not managed by the

---

[2] The organization uses Oracle PeopleSoft ®.

ERP. This makes the process inefficient, as only 32% of orders are delivered within 1 month, which is the user expected time.

The users (professors, research and administrative staff) frequently present complaints about delays and excessive paperwork in the process. Although professors must make purchases for research projects having 1 or 2 year time frames, the purchase of an imported good may take more than 6 months, which certainly impacts the schedule of these projects.

Process Scope and Stakeholder Identification

Figure 4 shows the process scope diagram, where it can be seen that the main input is the requisition made by departments and areas of the university. Said requisition starts a process that finishes when the product is delivered to the areas and the supplier has been paid. There is a procedure for the order approval subprocess, but there is no business rule specifying the maximum time allowed for this step. There is also a good governance code for managing suppliers and contracts. Enablers correspond to two different resources of the process: the information system (ERP system) and the staff involved in the process.

The shaded boxes in figure 4 represent the process stakeholders: departments, suppliers, purchasing board, and both the Administrative and IT offices.

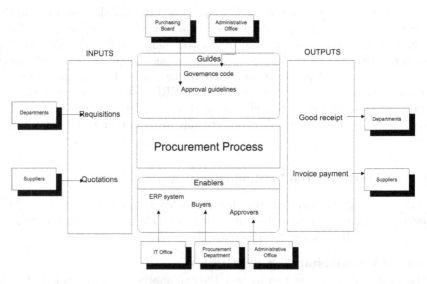

Fig. 4. Process scope analysis

Gap Analysis

The gap analysis was used to represent the current (as-is) and expected (to-be) process performances, as mediated by process redesign. In order to determine the expected performance, it is important to ask the stakeholders why the process should be improved and what the expected performance is in terms of key process indicators. Figure 5 shows the main performance and capability gaps and the tools that were used in the analysis and improvement of the process.

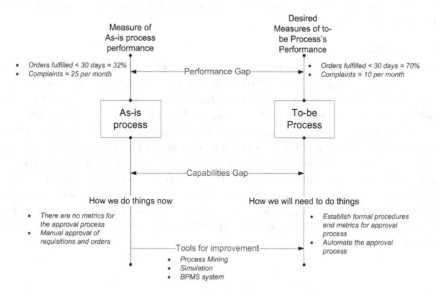

Fig. 5. Gap analysis

Definition of Project Goals

The desired amendments of the to-be process turn to be the project goals:

- Reducing cycle time to ensure that 70% of orders are delivered within 1 month.
- Reduce the number of user complaints.

B. Phase II: Analyzing Business Process (as-is)

This phase begins with event log data extraction in order to discover the real process model and to apply data mining techniques for an in-depth process analysis. The objective of this phase is to establish process improvement opportunities.

Event Log Extraction

In this phase, the event log is extracted from the ERP system. The information supplied by the log includes case id, time stamps, activities and performers (originators) of the procurement process. In addition, there is information regarding each order such as requested product or service, supplier, requesting department, cost, product family and the person approving each purchase requisition or order.

The original log contained one year of historical data, corresponding to 15,091 cases. The quality of the log was inspected in the statistical package[3] , which allowed finding some missing data and outliers. After cleaning the log, the cases were reduced to 8,987.

Process Discovery

Through the application of process mining algorithms such as alpha mining [14], heuristic mining [15] or genetic mining [16] it is possible to automatically discover the actual process model using the ProM software functionality. This model can be represented in a Petri Net, or through BPMN notation. Because of its mathematical foundations, process mining uses Petri Nets in most applications, which allows the implementation of analysis techniques [3]. Some case studies make use of Colored Petri Nets (CPN) because of their simulation capabilities in packages such as CPN tools [17]. In the current case, the alpha algorithm was used due to the low complexity of the process model and paths. For more complex processes, genetic algorithms or the heuristic mining algorithm are recommended. Figure 6 shows the studied procurement process modeled in a Petri Net.

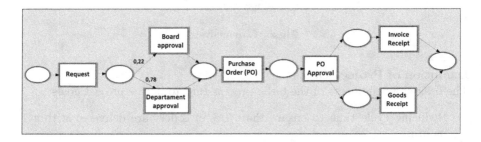

Fig. 6. Procurement process

Process Performance Analysis

The event log was assessed through descriptive statistics in order to analyze some key process indicators such as cycle time, cycle time per buyer and buyer productivity, among others. Figure 7 shows a box plot of cycle times per buyer, which exhibits a high variability in mean time cycles between buyers and, in some cases, high variability within buyers. This analysis suggests an influence of the buyer in time cycles. This influence is going to be analyzed in depth in the data mining analysis section.

---

[3] IBM SPSS® was used for the data mining analysis.

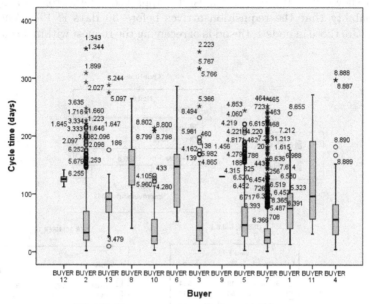

Fig. 7. Procurement cycle times per buyer

Table 4 presents some key findings of the process performance analysis.

Table 4. Key findings of the process performance analysis

The main bottleneck of the process is the purchase requisition approval subprocess
Mean cycle time is 50 days, with a standard deviation of 28 days. Only 32% of orders are delivered within 1 month.
Imports require thrice as much more time than local purchases. The minimum time required for an imported good is 40 days.
The mean cycle time per buyer is highly variable (Fig 7).

Data Mining Analysis

For a more detailed diagnosis of the purchase requisition approval subprocess, a decision tree analysis was made to discover the roles of the organization that delay the process. The database was split in three parts: training (40% of the records), validation (40% of the records) and test (20% of the records). The decision tree was growth and pruned using the classification and regression tree algorithm (CART) and Gini impurity. The CART procedure minimizes classification error given a tree size [18] Figure 8 shows the tree results in test data.

Figure 8 shows that if the order must be approved by the roles in node 2, the probability that the requisition arrives before 30 days is 1%. When the approvers are those in node 1, the odds of receiving the request within 30 days rise to 50%.

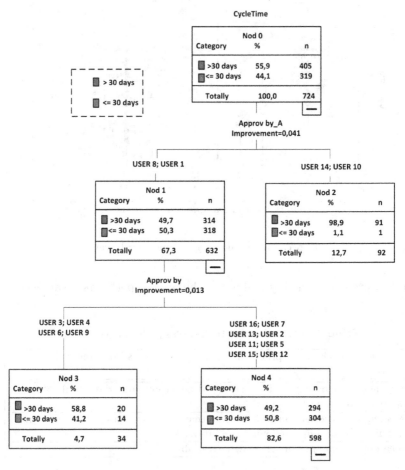

Fig. 8. Decision tree for purchase requisition approval

Table 5 presents the key findings of the purchase requisition approval decision tree analysis.

Table 5. Key findings of the purchase requisition approval decision tree analysis

The person that approves the purchase request has a significant impact on the probability of receiving the request within 30 days.

Root Cause Analysis

The *Cause and Effect* analysis was used to determine the cause of the problem. Through this tool, the roles involved in the execution of the process identified the major causes of delay in purchase requisition approval. Figure 9 shows these causes as classified by categories.

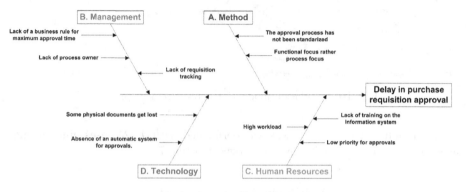

Fig. 9. Root cause analysis

Table 6 presents the key findings of this analysis.

Table 6. Key findings of this analysis

One of the main approval delay causes is that the physical documents that are handled in the process are not managed in a central repository.
Given that there is no business rule determining a time limit for approvals, approvers do not give the required priority to this process.

C. Phase III: Redesigning the Business Process (to-be)

In the redesign stage, the different process improvement options are simulated and evaluated.

Simulation Model

Based on the process discovered in phase 1, and on processing (P) and waiting times (W) calculated through the statistical analysis of the event log data, a simulation model was generated. Figure 10 shows the simulation model, which was obtained in the Process Modeler application.

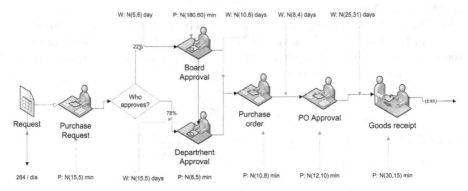

Fig. 10. Simulation model of the procurement process

Simulation of the as-is process was performed with 4,224 cases, finding and average cycle time of 50.72 days, as shown in Table 7.

Table 7. Average process cycle time

Scoreboard				
Scenario	Name	Total Exits	Average Time In System (Days)	Expected Average Time (Days)
Baseline	Request	4.224,00	50.72	30

Process Improvement Alternatives

The process improvement alternatives were defined to overcome the issues found in the as-is analysis phase. Said alternatives are shown in table 8.

Table 8. Process improvement alternatives

Process improvement alternative	
1. Removing the purchase order approval process.	Out of the 8,987 analyzed cases, no purchase order was rejected, so this control can be eliminated, the responsibility lying on the purchase requisition approval process.
2. Establishing an approval time limit business rule.	Said business rule would define that the approvers have a maximum of 5 days for purchase requisition approval.

Simulation and "what if" Analysis

The different improvement alternatives were simulated in corresponding scenarios.

– Scenario 1: Removal of the purchase order approval process.

- Scenario 2: Establishment of a business rule stating that approvers have a maximum of 5 days for purchase requisition approval.
- Scenario 3: Scenario 1 + Scenario 2

Table 9 shows the key findings of the simulation analysis

Table 9. Key findings of the simulation analysis

Table 9. Key findings of simulation and what if analysis		
Scenario 1: If the purchase order approval process was removed, the cycle time could be reduced to 43 days.		
Scoreboard		
Name	Total Exits	Average Time in System (Day)
Solicitud	4224,00	42,81
Scenario 2: If the university established a business rule specifying that approvers have a maximum of 5 days to approve purchase requisitions, the cycle time could be reduced to 43 days.		
Scoreboard		
Name	Total Exits	Average Time in System (Day)
Solicitud	4224,00	40,23
Scenario 3: Through the simultaneous implementation of scenarios 1 and 2, the cycle time could be reduced to 35 days.		
Scoreboard		
Name	Total Exits	Average Time in System (Day)
Solicitud	4224,00	34,76

D. Phase: Implementation
At this stage, the selected alternatives are implemented to improve the business process. For this case study, scenario 3 presented in Table 9 allows decreasing the cycle time to 35 minutes, thus constituting the alternative that is going to be recommended to the University for Implementation.

4.2   Lessons Learned

One of the key success factors for the implementation of the proposed methodology is the involvement of the people (users) playing a role in the actual execution of the business process. User knowledge is crucial for the interpretation of the cases, activities and variables of the process' event log, especially when it comes to preventing data misinterpretation and organizing a log that represents the actual execution of the process. Data extraction from, and cleansing of the event log is a crucial step that must be carried out in close connection with the users because they are the ones who know the real facts about outlier values and missing or wrong data.

The sequence proposed in this methodology does not necessarily have to be executed in that same order. Tools like process performance analysis, data mining and root cause analysis can be used in any order and may be complemented with other tools like Statistical Process Control from Six Sigma or Value Stream Mapping from Lean. These tools are complementary and might be useful in complex business processes where data mining and root cause analysis are not enough for a complete as-is process analysis.

Although the currently available process mining packages have been evolving in functionality, they still need to be more user-friendly, especially regarding data display techniques. Working with state-of-the-art algorithms, ProM is particularly useful for process discovery, but the resulting petri nets are not easy to interpret by the business user. Disco from Fluxicom is emerging as a user friendly package that provides more understandable visualization and animation tools. Providing adequate functionalities for finding missing data and outliers, SPSS and SAS are helpful and robust statistical packages when it comes to event log cleaning. Data mining analyses such as cluster and decision trees can be used with these applications.

5   Conclusions and Further Research

The present paper presents a methodological approach to process redesign that combines simulation techniques, data mining and process mining tools, as well as the tools of the understanding phase of the BPtrends methodology [10]. These tools and techniques are complementary to one another, and their integration contributes to achieving the goals set for each phase of the methodology.

BPtrend tools are useful for the understanding phase of the project, in which the scope of the process is established, the gap analysis between as-is and to-be is performed, and the stakeholders agree on the expected performance of the business process.

On the other hand, specific process mining techniques such as alpha, heuristics or genetic algorithms allow both discovering the actual process model and checking for compliance with business rules and procedures. Said model is used to construct the as-is simulation model.

Data mining techniques such as decision trees and cluster analysis are useful for determining the variables that influence process cycle times and to determine the odds of executing a process within a certain time limit.

Simulation benefits greatly from process mining since the latter provides the parameters that are needed to construct the simulation model, based on the real process model. Simulation makes it possible to test different process redesign alternatives before implementing them, thus becoming a valuable decision-making tool.

A process redesign project requires more than a single tool to achieve the expected results. Although process mining provides tools for process diagnosis and analysis, it must be complemented with other methodologies and techniques such as simulation and other process improvement tools that allow understanding and planning the process redesign effort.

The methodological approach proposed in this paper needs to be validated in other case studies reaching the implementation phase, in order to assess whether it meets the expected results. Further research is needed to determine how the event log source (ERP, WFMS, CRM) determines the necessary log extraction, transformation and cleansing activities.

References

1. van der Aalst, W.M.P., et al.: Process mining manifesto. In: Daniel, F., Barkaoui, K., Dustdar, S. (eds.) BPM Workshops 2011, Part I. LNBIP, vol. 99, pp. 169–194. Springer, Heidelberg (2012)
2. Van der Aalst, W.M.P.: Process Mining: Discovery, Conformance and Enhancement of Business Process. Springer, Berlin (2011)
3. Stahl, C.: Modeling Business Process: A Petri Net-Oriented Approach. The MIT Press, Cambridge (2011)
4. Rozinat, A., Mans, R., Song, M., Van der Aalst, W.M.P.: Discovering colored petri nets from event logs. International Journal of Software Tools for Technology Transfer 10(1), 57–74 (2007)
5. Jans, M., Van der Alast, W.M.P., Werf, J., Lybaert, N., Vanhoof, K.: A business process mining application for internal transaction fraud mitigation. Expert Systems with Applications 38(3), 13351–13359 (2011)
6. Van der Aalst, W.M.P., et al.: Business process mining: an industrial application. Information Systems 32(5), 713–732 (2007)
7. Van der Aalst, W.M.P., Schonenberg, M., Song, M.: Time prediction based on process mining. Information Systems 36(2), 450–475 (2011)
8. Aguirre, S.: Aplicacion de mineria de procesos al proceso de compras de la puj (2011) (unpublished)
9. Weske, M.: Business Process Management: Concepts, Languages, Architectures, Berlin (2010)
10. Harmon, P., Davenport, T.: Business Process Change. Morgan Kaufmann, Burlington (2007)
11. Bozkaya, M., Gabriels, J., Werf, J.: Process diagnostics: A method based on process mining, information, process, and knowledge management. In: International Conference on eKNOW 2009, pp. 22–27 (2009)
12. Rebuge, A., Ferreira, D.: Business process analysis in healthcare environments: A methodology based on process mining. Information Systems 37(2), 99–116 (2012)
13. Maruster, L., Van Beest, L.: Redesigning business processes: a methodology based on simulation and process mining techniques. Knowledge and Information Systems 21(3), 267–297 (2009)
14. Van der Aalst, W.M.P., Weijters, A.J.M.M., Maruster, L.: Workflow mining: discovering process models from event logs. IEEE Transactions on Knowledge and Data Engineering 16(9), 1128–1142 (2004)
15. Weijters, A.J.M.M., Ribeiro, J.: Flexible heuristics miner. BETA working Paper Series. Eindhoven University of Technology, Eindhoven (2010)
16. Medeiros, A., Weijters, A.J.M.M., Van der Aalst, W.M.P.: Genetic process mining: An experimental evaluation. Data Mining and Knowledge Discover 14(2), 245–304 (2007)
17. Rozinat, A., Mans, R., Song, M.: Discovering simulation models. Information Systems 34(3), 305–327 (2009)
18. Shmueli, G., Patel, N., Bruce, P.: Data mining for business intelligence, Hoboken, NJ, USA (2007)

Improving Business Process Models Using Observed Behavior

J.C.A.M. Buijs[1,2], M. La Rosa[2,3], H.A. Reijers[1],
B.F. van Dongen[1], and W.M.P. van der Aalst[1]

[1] Eindhoven University of Technology, The Netherlands
{j.c.a.m.buijs,h.a.reijers,b.f.v.dongen,w.m.p.v.d.aalst}@tue.nl
[2] Queensland University of Technology, Australia
m.larosa@qut.edu.au
[3] NICTA Queensland Research Lab, Australia

Abstract. Process-aware information systems (PAISs) can be configured using a *reference* process model, which is typically obtained via expert interviews. Over time, however, contextual factors and system requirements may cause the operational process to start deviating from this reference model. While a reference model should ideally be updated to remain aligned with such changes, this is a costly and often neglected activity. We present a new process mining technique that automatically improves the reference model on the basis of the observed behavior as recorded in the event logs of a PAIS. We discuss how to balance the four basic quality dimensions for process mining (fitness, precision, simplicity and generalization) and a new dimension, namely the structural similarity between the reference model and the discovered model. We demonstrate the applicability of this technique using a real-life scenario from a Dutch municipality.

1 Introduction

Within the area of *process mining* several algorithms are available to automatically discover process models. By only considering an organization's records of its operational processes, models can be derived that accurately describe the operational business processes. Organizations often use a *reference* process model, obtained via expert interviews, to initially configure a process. During execution however the operational process typically starts deviating from this reference model, for example, due to new regulations that have not been incorporated into the reference model yet, or simply because the reference model is not accurate enough.

Process mining techniques can identify where reality deviates from the original reference model and especially how the latter can be adapted to better fit reality. Not updating the reference model to reflect new or changed behavior has several disadvantages. First of all, such a practice will overtime drastically diminish the reference model's value in providing a factual, recognizable view on how work

P. Cudre-Mauroux, P. Ceravolo, and D. Gašević (Eds.): SIMPDA 2012, LNBIP 162, pp. 44–59, 2013.

is accomplished within an organization. Second, a misaligned reference model cannot be used to provide operational support in the form of, e.g., predictions or recommendations during the execution of a business process.

A straightforward approach to fix the misalignment between a reference model and reality is to simply discover a new process model from scratch, using automated process discovery techniques from process mining [1]. The resulting model may reflect reality better but may also be very different from the initial reference model. Business analysts, process owners and other process stakeholders, may heavily rely on the initial reference model to understand how a particular process functions. Confronting them with an entirely new model may make it difficult for them to recognize its original, familiar ingredients and understand the changes in the actual situation. As a result, a freshly discovered process model may actually be useless in practice.

In this paper, we propose to use process mining to discover a process model that accurately describes an existing process yet is *very similar* to the initial reference process model. To explain our approach, it is useful to reflect on the four basic quality dimensions of the process model with respect to the observed behavior [1,2] (cf. Figure 1a). The *replay fitness* dimension quantifies the extent to which the discovered model can accurately replay the cases recorded in the log. The *precision* dimension measures whether the discovered model prohibits behavior which is not seen in the event log. The *generalization* dimension assesses the extent to which the resulting model will be able to reproduce possible future, yet unseen, behavior of the process. The complexity of the discovery process model is captured by the *simplicity* dimension, which operationalizes Occam's Razor.

Following up on the idea to use process mining for aligning reference process models to observed behaviors, we propose to add a fifth quality dimension to this spectrum: *similarity* to a given process model. By incorporating this dimension, we can present a discovered model that maximizes the four dimensions while remaining aligned, as far as possible, with the intuitions and familiar notions modeled in a reference model.

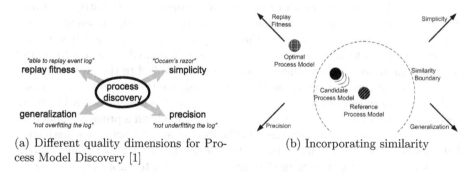

(a) Different quality dimensions for Process Model Discovery [1]

(b) Incorporating similarity

Fig. 1. Adding similarity as a process model quality dimension

Figure 1b illustrates the effects of introducing this additional dimension. By setting a *similarity boundary*, the search for a model that balances the initial four quality dimensions is restrained. In this way, a new version of the reference model can be found that is similar to the initial reference model yet is improved with respect to its fit with actual behavior. Clearly, if the similarity boundary is being relaxed sufficiently (i.e. the discovered model is allowed to deviate strongly from the reference model), it is possible to discover the *optimal* process model. Such an optimal model, as explained, may not be desirable to use for process analysts and end users as a reference point, since they may find it difficult to recognize the original process set-up within it.

The remainder of the paper is structured as follows. In Section 2 we present related work in the area of process model improvement and process model repair. In Section 3 we present our approach using a genetic algorithm to balance the different quality dimensions while in Section 4 we show how to incorporate the similarity dimension in our approach. In Section 5 we show the results of applying our technique to a small example. In Section 6 the technique is applied to a real life case. Finally, Section 7 concludes the paper.

2 Related Work

Automatically improving or correcting process models using different sources of information is an active research area. Li et. al. [15] discuss how a reference process model can be discovered from a collection of process model variants. In their heuristic approach they consider the structural distance of the discovered reference model to the original reference model as well as the structural distance to the process variants. By balancing these two forces they make certain changes to the original reference model to make it more similar to the collection of process model variants. Compared to our approach, here the starting point is a collection of process variants, rather than a log.

An approach aimed to automatically correct errors in an *unsound* process model (a process model affected by behavioral anomalies) is presented by Gambini et. al. [11]. Their approach considers three dimensions: the structural distance, behavioral distance and 'badness' of a solution w.r.t. the unsound process model, whereby 'badness' indicates the ability of a solution to produce traces that lead to unsound behavior. The approach uses simulated annealing to simultaneously minimize all three dimensions. The edits applied to the process model are aimed to correct the model rather than to balance the five different forces.

Detecting deviations of a process model from the observed behavior has been researched, among others, by Adriansyah et. al. [2,4]. Given a process model and an event log, deviations are expressed in the form of *skipped* activities (activities that should be performed according to the model, but do not occur in the log) and *inserted* activities (activities that are not supposed to happen according to the model, but that occur in the log). A cost is attributed to these operations based on the particular activity being skipped/inserted. Based on this information an alignment can be computed between the process model and the log, which indicates how well the process model can describe the recorded behavior. While this approach provides an effective measure for the replay fitness quality dimension of Figure 1a, the approach per se does not suggest any corrections to rectify the process model's behavior.

The work of Fahland et. al. [10] provides a first attempt at repairing process models based on observed behavior. In their notion, a process model needs repair if the observed behavior cannot be replayed by the process model. This is detected using the alignment between the process model and the observed behavior of [2, 4]. The detected deviations are then repaired by extending the process model with sub-processes nested in a loop block. These fixes are applied repeatedly until a process model is obtained that can perfectly replay the observed behavior. This approach extends the original process model's behavior by adding new fragments that enable the model to replay the observed behavior (no existing fragments are removed). The main disadvantage of this approach is that only one aspect of deviation, namely that of not being able to replay the observed behavior, is considered. Moreover, since repairs add transitions to the model, by definition, the model can only become more complex and less precise. It is unclear how to balance all five quality dimensions by extending the work in [10].

3   Our Mining Technique

In this section we briefly introduce our flexible evolutionary algorithm first presented in [6]. This algorithm can seamlessly balance four process model quality dimensions during process discovery.

3.1   Process Trees

Our approach internally uses a tree structure to represent process models. Because of this, we only consider sound process models. This drastically reduces the search space thus improving the performance of the algorithm. Moreover, we can apply standard tree change operations on the process trees to evolve them further, such as adding, removing and updating nodes.

Figure 2 shows the possible operators of a process tree and their translation to a Petri net. A process tree contains operator nodes and leaf nodes. An operator node specifies the relation between its children. Possible operators are sequence ($\rightarrow$), parallel execution ($\wedge$), exclusive choice ($\times$), non-exclusive choice ($\vee$) and loop execution ($\circlearrowleft$). The order of the children matters for the sequence and loop operators. The order of the children of a sequence operator specifies the order in which the children are executed (from left to right). For a loop, the left child is the 'do' part of the loop. After the execution of this part the right child, the 'redo' part, might be executed. After this execution the 'do' part is again enabled. The loop in Figure 2 for instance is able to produce the traces $\langle A \rangle$, $\langle A, B, A \rangle$, $\langle A, B, A, B, A \rangle$ and so on. Existing process models can be translated to the process tree notation, possibly by duplicating activities.

3.2   Quality Dimensions

To measure the quality of a process tree, we consider one metric for each of the four quality dimensions, as we proposed in [6]. We base these metrics on existing work in each of the four areas [2, 4] and we adapt them for process trees, as discussed below. For the formalization of these metrics on process trees we refer to [6].

Replay fitness quantifies the extent to which the model can reproduce the traces recorded in the log. We use an alignment-based fitness computation defined in [4] to compute the fitness of a process tree. Basically, this technique aligns as many events as possible from the trace with activities in an execution of the model (this results in a so-called *alignment*). If necessary, events are skipped, or activities are inserted without a corresponding event present in the log. Penalties are given for skipping and inserting activities. The total costs for the penalties are then normalized, using information on the maximum possible costs for this event log and process model combination, to obtain a value between 1 (perfect) and 0 (bad).

Precision compares the state space of the tree execution while replaying the log. Our metric is inspired by [5] and counts so-called escaping edges, i.e. decisions that are possible in the model, but never made in the log. If there are no escaping edges, the precision is perfect. We obtain the part of the state space used from information provided by the replay fitness, where we ignore events that are in the log, but do not correspond to an activity in the model according to the alignment.

Generalization considers the frequency with which each node in the tree needs to be visited if the model is to produce the given log. For this we use the alignment provided by the replay fitness. If a node is visited more often, then we are more certain that its behavior is (in)correct. If some parts of the tree are very infrequently visited, generalization is bad.

Simplicity quantifies the complexity of the model. Simplicity is measured by comparing the size of the tree with the number of activities in the log. This is based on the finding that the size of a process model is the main factor

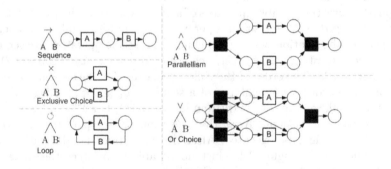

Fig. 2. Relation between process trees and block-structured Petri nets

for perceived complexity and introduction of errors in process models [16]. Furthermore, since we internally use binary trees, the number of leaves of the process tree has a direct influence on the number of operator nodes. Thus, the tree in which each activity is represented exactly once is considered to be as simple as possible.

The four metrics above are computed on a scale from 0 to 1, where 1 is optimal. Replay fitness, simplicity and precision can reach 1 as optimal value. Generalization can only reach 1 in the limit i.e., the more frequent the nodes are visited, the closer the value gets to 1. The flexibility required to find a process model that optimizes a weighted sum over the four metrics can efficiently be implemented using a genetic algorithm.

3.3 The ETM Algorithm

In order to be able to seamlessly balance the different quality dimensions we implemented the ETM algorithm (which stands for *Evolutionary Tree Miner*). In general, this genetic algorithm follows the process shown in Figure 3. The input of the algorithm is an event log describing the observed behavior and, optionally, one or more reference process models. First, the different quality dimensions for each candidate currently in the population are calculated, and using the weight given to each dimension, the *overall fitness* of the process tree is calculated. In the next step certain stop criteria are tested such as finding a tree with the desired overall fitness, or exceeding a time limit. If none of the stop criteria are satisfied, the candidates in the population are changed and the fitness is again calculated. This is continued until at least one stop criterion is satisfied and the best candidate (highest overall fitness) is then returned.

The genetic algorithm has been implemented as a plug-in for the ProM framework [18]. We used this implementation for all experiments presented in this paper. The algorithm stops after $1,000$ generations or sooner if a candidate with perfect overall fitness is found before. In [7] we empirically showed that $1,000$ generations are typically enough to find the optimal solution, especially

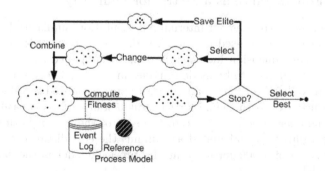

Fig. 3. The different phases of the genetic algorithm

for processes with few activities. All other settings were selected according to the optimal values presented in [7].

4   Similarity as the 5th Dimension

In order to extend our ETM algorithm for process model improvement we need to add a metric to measure the similarity of the candidate process model to the reference process model. Similarity of business process models is an active area of research [3, 8, 9, 12–15, 19]. We distinguish two types of similarity: i) *behavioral similarity* and ii) *structural similarity*. Approaches focusing on behavioral similarity, e.g. [3, 8, 9, 13, 19], encode the behavior described in the two process models to compare using different relations. Examples are causal footprints [9], transition adjacency relations [19], or behavioral profiles [13]. By comparing two process models using such relations, it is possible to quantify behavioral similarity in different ways.

Approaches focusing on structural similarity only consider the graph structure of models and abstract from the actual behavior, e.g., heuristic approaches like [15], only focus on the number of common activities ignoring the connecting arcs, or vice versa, ignore the actual activities to only consider the arcs. Most approaches [8, 12, 14] provide a similarity metric based on the minimal number of edit operations required to transform one model into the other model, where an edit is either a node or an arc insertion/removal.

Both behavioral and structural similarity approaches first require a suitable *mapping* of nodes between the two models. This mapping can be best achieved by combining techniques for syntactic similarity (e.g. using string-edit distance) with techniques for linguistic similarity (e.g. using synonyms) [8].

Our algorithm only needs to consider the structural similarity, since the event log already captures the behavior that the process model should describe. Recall that the behavior of the reference model w.r.t. the logs is already measured by means of the four mining dimensions (Fig. 3.2). Hence, we use structural similarity to quantify the fifth dimension.

4.1   Tree Edit Distance as a Metric for Similarity

Since we use process trees as our internal representation, similarity between two process trees can be expressed by the tree edit distance for ordered trees. The tree edit distance dimension indicates how many simple edit operations (add, remove and change) need to be made to nodes in one tree in order to obtain the other tree. Since the other four quality metrics are normalized to values between 0 and 1, we need to do the same for the edit distance. This is easily done by making the number of edits relative to the sum of the size of both trees. The similarity score finally is calculated as 1 minus the edit distance ratio. Hence, a similarity score of 1.000 means that the process model is the same as the reference model.

Figure 4 shows examples of each of the three edit operations. The reference tree is shown in Figure 4a. Figure 4b shows the result after deleting activity B from the tree. Our trees are binary trees, meaning that each non-leaf node has exactly 2 children. Therefore, the × operator node is also removed. The removal of B from the tree results in an edit distance of 2. The similarity is $1 - \frac{2}{5+3} = 0.75$.

The process tree shown in Figure 4c has activity D added in parallel to activity A. This also results in 2 edits since a new ∧ operator node needs to be added, including a leaf for activity D. Since the resulting tree has grown, the relative edit distance is less than when part of the tree is deleted. Finally, changing a node as shown in Figure 4d, where the root → operator is changed into an ∧ operator, only requires 1 edit operation.

We use the Robust Tree Edit Distance (RTED) algorithm [17] to calculate the edit distance between two ordered trees. The RTED approach first computes the optimal strategy to use for calculating the edit distance. It then calculates the edit distance using that strategy. Since the overhead of determining the optimal strategy is minimal, this ensures best performance and memory consumption, especially for larger trees. However, it is important to realize that our approach is not limited to the RTED algorithm. Furthermore, although in this paper we only consider a single distance metric, it is possible to incorporate multiple metrics (for example looking at both structural and behavioral similarity).

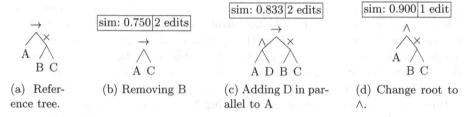

(a) Reference tree.   (b) Removing B   (c) Adding D in parallel to A   (d) Change root to ∧.

Fig. 4. Examples of possible edits on a tree (a) and respective similarities

Table 1. The event log

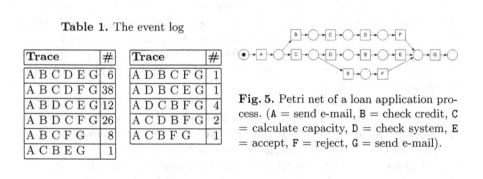

Trace	#	Trace	#
A B C D E G	6	A D B C F G	1
A B C D F G	38	A D B C E G	1
A B D C E G	12	A D C B F G	4
A B D C F G	26	A C D B F G	2
A B C F G	8	A C B F G	1
A C B E G	1		

Fig. 5. Petri net of a loan application process. (A = send e-mail, B = check credit, C = calculate capacity, D = check system, E = accept, F = reject, G = send e-mail).

Table 2. Different weight combinations and the resulting fitness values for the simple example

Weights					Quality					
Sim	f	p	g	s	Sim	edits	f	p	g	s
100	10	1	1	1	**1.000**	0	0.880	1.000	0.668	0.737
10	10	1	1	1	**0.935**	3	1.000	0.885	0.851	0.737
1	10	1	1	1	**0.667**	12	1.000	0.912	0.889	1.000
0.1	10	1	1	1	**0.639**	13	1.000	0.923	0.889	1.000
10	**0**	1	1	1	1.000	0	**0.880**	1.000	0.668	0.737
10	10	**0**	1	1	0.935	3	1.000	**0.849**	0.851	0.737
10	10	1	**0**	1	0.978	1	0.951	0.992	**0.632**	0.737
10	10	1	1	**0**	0.935	3	1.000	0.885	0.851	**0.737**

5   Experimental Evaluation

Throughout this section we use a small example to explain the application and use of our approach. Figure 5 describes a simple loan application process of a financial institute which provides small consumer credits through a webpage. The figure shows the process as it is known within the company. When a potential customer fills in a form and submits the request from the website, the process starts by executing activity A which notifies the customer with the receipt of the request. Next, according to the process model, there are two ways to proceed. The first option is to start with checking the credit (activity B) followed by calculating the capacity (activity C), checking the system (activity D) and rejecting the application by executing activity F. The other option is to start with calculating the capacity (activity C) after which another choice is possible. If the credit is checked (activity B) then finally the application is rejected (activity F). Another option is the only one resulting in executing E, concerned with accepting the application. Here activity D follows activity C, after which activity B is executed, and finally activity E follows. In all three cases the process ends with activity G, which notifies the customer of the decision made.

However, the observed behavior, as is recorded in the event log shown in Table 1, deviates from this process model. The event log contains 11 different traces whereas the original process model only allows for 3 traces, i.e., modeled and observed behavior differ markedly. To demonstrate the effects of incorporating the similarity between process trees, we run the extended ETM algorithm on the example data of Table 1.

In [6] we showed that, on this data set, the optimal weights are 10 for replay fitness and 1 for precision, generalization and simplicity. In the first experiment (Section 5.1), we only change the similarity weight to vary the amount of change we allow. In the second experiment (Section 5.2) we fix the weight for similarity and ignore each of the other four quality dimensions, one at a time. The experiment settings and their results are shown in Table 2.

5.1   Varying the Similarity Weight

Figure 6a shows the process tree that is discovered when giving the similarity a weight of 100. The similarity ratio is 1.000, indicating that no change has taken place. Apparently no change in the tree would improve the other dimensions enough to be beneficial.

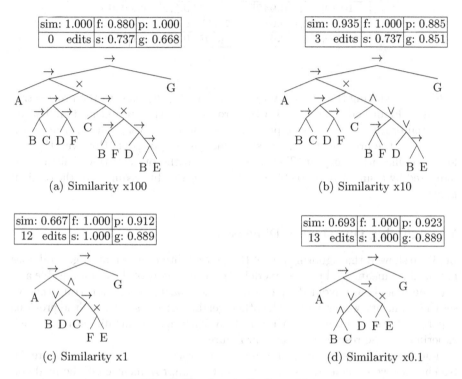

(a) Similarity x100

(b) Similarity x10

(c) Similarity x1

(d) Similarity x0.1

Fig. 6. Varying similarity weight

If we reduce the similarity weight to 10, the process tree as shown in Figure 6b is discovered. Three edits have been applied: in the bottom-right part of the tree two → and an × operator have been changed to ∧ and ∨. This allows for more behavior, as is indicated by the increase in replay fitness of 0.220. Also, generalization increased by 0.183, at the cost of a decrease in precision of 0.115.

If we lower the weight of the similarity to 1, we get the process tree as shown in Figure 6c. This process tree requires 12 edits starting from the original tree and is very different from the process tree we started with. However, compared to the previous process tree, the other 4 quality dimensions have improved overall. Replay fitness has now reached a value of 1.000 since this process tree allows skipping activity D. Also, simplicity reached 1.000 since no activities are duplicated or missing.

Table 3. Different weight combinations and the resulting fitness values for the practice application

Weights					Quality					
Sim	f	p	g	s	Sim	edits	f	p	g	s
1000	10	1	1	1	1.000	0	0.744	0.785	0.528	0.755
100	10	1	1	1	0.990	1	0.858	0.799	0.566	0.792
10	10	1	1	1	0.942	6	0.960	0.770	0.685	0.815
1	10	1	1	1	0.650	42	0.974	0.933	0.747	0.613
0.1	10	1	1	1	0.447	83	0.977	0.862	0.721	0.519

Finally, reducing the similarity weight to 0.1 provides us with the process tree shown in Figure 6d, which is also the process tree that would be found when no initial process tree has been provided, i.e., pure discovery. The only improvement w.r.t. the previous tree is the slight increase in precision. However, the tree looks significantly different. The resemblance to the original tree is little as is indicated by a similarity of 0.693, caused by the 13 edits required to the original model.

5.2 Ignoring One Quality Dimension

In [6] we showed that ignoring one of the four quality dimensions in general does not produce meaningful process models. However, many of these undesirable and extreme models are avoided by considering similarity. To demonstrate this we set the similarity weight to 10. The other weights are the same as in the previous experiment: 10 for fitness, 1 for the rest. We then ignore one dimension in each experiment. The results are shown in Figure 7.

Ignoring the fitness dimension results in the process tree as shown in Figure 7a. No changes were made, demonstrating that no improvement could be made on the other three dimensions that was worth the edit.

If precision is ignored, the result is the process tree as shown in Figure 7b. Replay fitness and generalization improved by applying 3 edits. The tree of Figure 6b, where we used the same similarity weight but included precision, only 1 edit was allowed. By removing the restriction on precision, it is worth to apply more edits to improve replay fitness and generalization.

We do not see this effect as strongly when we ignore generalization or simplicity. The resulting process trees, shown in Figure 7c and Figure 7d, are very similar to the original one with only 1 edit.

This experiment shows that considering similarity to a reference process model avoids the extreme cases encountered in [6].

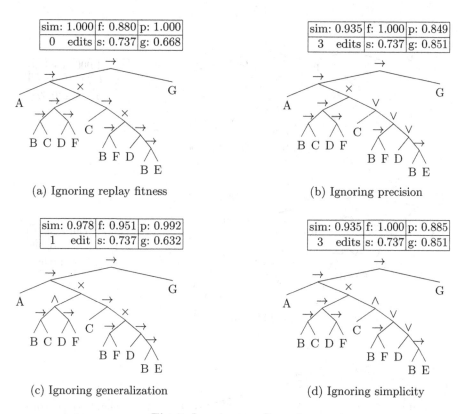

(a) Ignoring replay fitness   (b) Ignoring precision

(c) Ignoring generalization   (d) Ignoring simplicity

Fig. 7. Ignoring one dimension

6   Application in Practice

Within the context of the CoSeLoG project, we are collaborating with ten Dutch municipalities that are facing the problem addressed in this paper.[1] The municipalities have implemented case management support, using a particular reference model. Now they are deriving new, possibly shared, reference models because they want to align each model with their own real process and the real processes in the municipalities they are collaborating with.

One of the municipalities participating in the CoSeLoG project recently started looking at one of their permit processes. The reference model used in the implementation was very detailed, with many checks that the employees in practice did not always do (usually with good reasons). Therefore, they were interested in a model that looks similar to the original reference model, but still shows most of the behavior actually observed. For this we applied our technique to discover different variants of the process model, focusing on different quality combinations, while maintaining the desired similarity to the reference model. For this

---

[1] See http://www.win.tue.nl/coselog

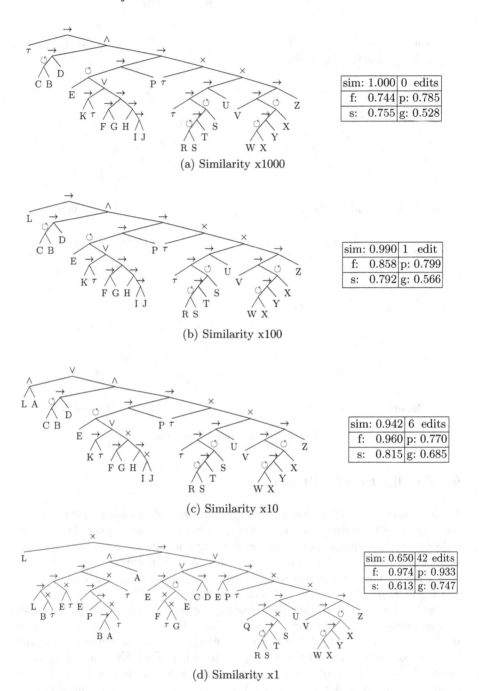

(a) Similarity x1000

(b) Similarity x100

(c) Similarity x10

(d) Similarity x1

Fig. 8. Process trees for the municipality process

application we only used the first part of the process which contains a total of 27 different activity labels, which we anonymized using letters from A to Z plus AA.

The experiment settings and their results are shown in Table 3. We experimented with fixed weights for the original four quality dimensions, and only changed the weight for the similarity to the reference model. The results confirm the intuition that reducing the weight of the similarity dimension allows more edits to be made (cf. the quality part of the table). In general, we also see that by allowing more edits, most of the quality dimensions improve. Of course, there is a trade-off between different dimensions. Since we weight the replay fitness dimensions 10 times more than the others, we see that this dimension always improves, sometimes at the cost of the other dimensions.

Figure 8a shows the process tree that is discovered using a similarity weight of 1,000. This is the same process tree as created from the process model provided by the municipality. All four quality dimensions are relatively bad and many improvements are possible. However, none of these improvements were applied since one change needs to drastically improve the process tree to be worth it.

If we set the similarity weight to 100 we obtain the process tree of Figure 8b. Here one edit has been made, namely the left-most activity leaf node has been changed from τ to L. This single edit causes all four quality dimensions to improve, especially replay fitness. The original process model used a significantly different activity name than the one present in the event log, which was translated to a τ in the process tree.

If we further relax the importance of similarity by using a weight of 10, we obtain the process tree of Figure 8c. Here 6 edits have been made from the original model. The root node now changed to an $\vee$ to allow more behavior. The left branch of the root node also changed to allow more behavior, better suiting the recorded behavior. Also the operator node of activities I and J changed as well as the operator of their grandparent node. It appears that the event log contains a lot of short traces, only containing activities from the first part of the process. Some traces even contain activity L only. All the dimensions have improved after these changes, except precision which has slightly decreased.

Further relaxing the similarity we obtain the process trees of Figure 8d (weight of 1 – 42 changes) and Figure 8e (weight of 0.1 – 83 changes). Both these models have little to do with the original reference model. At the same time, the quality of these two process trees with respect to the log did not improve much, while their appearance did. Therefore, for this experiment, we propose the process tree of Figure 8c as the improved version of the reference model. By applying only

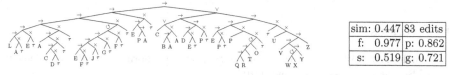

sim:	0.447	83 edits
f:	0.977	p: 0.862
s:	0.519	g: 0.721

(e) Similarity x0.1

Fig. 8. Process trees for the municipality process (cont'ed)

6 edits the process model has improved significantly, mainly on replay fitness (from 0.744 to 0.960), while still showing a great resemblance to the original reference model.

7   Conclusion

In this paper, we proposed a novel process mining algorithm that improves a given *reference* process model using observed behavior, as extracted from the event logs of an information system. A distinguishing feature of the algorithm is that it takes into account the structural similarity between the discovered process model and the initial reference process model. The proposed algorithm is able to improve the model with respect to the four basic quality aspects (fitness, precision, generalization and simplicity) while remaining as similar as possible to the original reference model (5th dimension). The relative weights of all five dimensions can be configured by the user, thus guiding the discovery/modification procedure. We demonstrated the feasibility of the algorithm through various experiments and illustrated its practical use within the CoSeLoG project.

A limitation of this paper is that it assumes that the deviations discovered from the logs are always rightful. Indeed, some process deviations do reflect an evolving business process due to new acceptable practices or regulations, and as such should be accommodated into the reference model. However, some other deviations may be the result of non-compliance or be caused by a sub-efficient execution. These undesirable deviations should be isolated and discarded in order to prevent bad practices form becoming a part of the reference model. In future work, we plan to implement a more fine-grained control on the different costs for edit actions on different parts of the process model. For example, edits on operators may have lower costs than edits on labels. In this way we can for instance restrict our changes to extensions of the original reference model, and prevent existing parts of the model from being changed. Also, pre-defined domain-specific constraints which the process model should adhere to can be fixed in this way. However, while these techniques may help produce better results, the identified deviations still need to be validated by a domain expert before making their way into the reference model. Only in this way we can ensure that false positives are properly identified.

Finally, we plan to conduct an empirical evaluation of the understandability of the process models discovered using our algorithm, as perceived by domain experts, and compare the results with those obtained with other process mining algorithms, which ignore the similarity dimension.

References

1. van der Aalst, W.M.P.: Process Mining: Discovery, Conformance and Enhancement of Business Processes. Springer (2011)
2. van der Aalst, W.M.P., Adriansyah, A., van Dongen, B.: Replaying History on Process Models for Conformance Checking and Performance Analysis. WIREs Data Mining and Knowledge Discovery 2(2), 182–192 (2012)

3. van der Aalst, W.M.P., de Medeiros, A.K.A., Weijters, A.J.M.M.: Process Equivalence: Comparing Two Process Models Based on Observed Behavior. In: Dustdar, S., Fiadeiro, J.L., Sheth, A.P. (eds.) BPM 2006. LNCS, vol. 4102, pp. 129–144. Springer, Heidelberg (2006)

4. Adriansyah, A., van Dongen, B., van der Aalst, W.M.P.: Conformance Checking using Cost-Based Fitness Analysis. In: Proceedings of EDOC, pp. 55–64. IEEE Computer Society (2011)

5. Adriansyah, A., Munoz-Gama, J., Carmona, J., van Dongen, B.F., van der Aalst, W.M.P.: Alignment Based Precision Checking. In: La Rosa, M., Soffer, P. (eds.) BPM Workshops 2012. LNBIP, vol. 132, pp. 137–149. Springer, Heidelberg (2013)

6. Buijs, J.C.A.M., van Dongen, B.F., van der Aalst, W.M.P.: On the Role of Fitness, Precision, Generalization and Simplicity in Process Discovery. In: Meersman, R., et al. (eds.) OTM 2012, Part I. LNCS, vol. 7565, pp. 305–322. Springer, Heidelberg (2012)

7. Buijs, J.C.A.M., van Dongen, B.F., van der Aalst, W.M.P.: A Genetic Algorithm for Discovering Process Trees. In: IEEE Congress on Evolutionary Computation, pp. 1–8. IEEE (2012)

8. Dijkman, R.M., Dumas, M., van Dongen, B.F., Käärik, R., Mendling, J.: Similarity of Business Process Models: Metrics and Evaluation. Information Systems 36(2), 498–516 (2011)

9. van Dongen, B.F., Dijkman, R., Mendling, J.: Measuring Similarity between Business Process Models. In: Bellahsène, Z., Léonard, M. (eds.) CAiSE 2008. LNCS, vol. 5074, pp. 450–464. Springer, Heidelberg (2008)

10. Fahland, D., van der Aalst, W.M.P.: Repairing Process Models to Reflect Reality. In: Barros, A., Gal, A., Kindler, E. (eds.) BPM 2012. LNCS, vol. 7481, pp. 229–245. Springer, Heidelberg (2012)

11. Gambini, M., La Rosa, M., Migliorini, S., Ter Hofstede, A.H.M.: Automated Error Correction of Business Process Models. In: Rinderle-Ma, S., Toumani, F., Wolf, K. (eds.) BPM 2011. LNCS, vol. 6896, pp. 148–165. Springer, Heidelberg (2011)

12. Jin, T., Wang, J., Wen, L.: Efficient retrieval of similar business process models based on structure. In: Meersman, R., et al. (eds.) OTM 2011, Part I. LNCS, vol. 7044, pp. 56–63. Springer, Heidelberg (2011)

13. Kunze, M., Weidlich, M., Weske, M.: Behavioral Similarity – A Proper Metric. In: Rinderle-Ma, S., Toumani, F., Wolf, K. (eds.) BPM 2011. LNCS, vol. 6896, pp. 166–181. Springer, Heidelberg (2011)

14. La Rosa, M., Dumas, M., Uba, R., Dijkman, R.: Business Process Model Merging: An Approach to Business Process Consolidation. ACM Transactions on Software Engineering and Methodology 22(2) (2013)

15. Li, C., Reichert, M., Wombacher, A.: The minadept clustering approach for discovering reference process models out of process variants. IJCIS 19(3-4), 159–203 (2010)

16. Mendling, J., Verbeek, H.M.W., van Dongen, B.F., van der Aalst, W.M.P., Neumann, G.: Detection and Prediction of Errors in EPCs of the SAP Reference Model. Data and Knowledge Engineering 64(1), 312–329 (2008)

17. Pawlik, M., Augsten, N.: RTED: A Robust Algorithm for the Tree Edit Distance. CoRR, abs/1201.0230 (2012)

18. Verbeek, H.M.W., Buijs, J.C.A.M., van Dongen, B.F., van der Aalst, W.M.P.: XES, XESame, and ProM 6. In: Soffer, P., Proper, E. (eds.) CAiSE Forum 2010. LNBIP, vol. 72, pp. 60–75. Springer, Heidelberg (2011)

19. Zha, H., Wang, J., Wen, L., Wang, C., Sun, J.: A Workflow Net Similarity Measure based on Transition Adjacency Relations. Computers in Industry 61(5), 463–471 (2010)

Process Prediction in Noisy Data Sets: A Case Study in a Dutch Hospital

Sjoerd van der Spoel, Maurice van Keulen, and Chintan Amrit

University of Twente, The Netherlands
{s.j.vanderspoel,m.vankeulen,c.amrit}@utwente.nl

Abstract. Predicting the amount of money that can be claimed is critical to the effective running of an Hospital. In this paper we describe a case study of a Dutch Hospital where we use process mining to predict the cash flow of the Hospital. In order to predict the cost of a treatment, we use different data mining techniques to predict the sequence of treatments administered, the duration and the final "care product" or diagnosis of the patient. While performing the data analysis we encountered three specific kinds of noise that we call *sequence noise*, *human noise* and *duration noise*. Studies in the past have discussed ways to reduce the noise in process data. However, it is not very clear what effect the noise has to different kinds of process analysis. In this paper we describe the combined effect of *sequence noise*, *human noise* and *duration noise* on the analysis of process data, by comparing the performance of several mining techniques on the data.

Keywords: process prediction, process mining, classification, cash flow prediction, data noise, case study.

1 Introduction

In the Netherlands, insurance companies play an important role in settling the finances of medical care. Hospitals and other care providers claim their costs for treating a patient with the patient's insurance company, who then bill their customer if there are costs not covered by the insurance.

Starting January 1st, 2012, Dutch hospitals are required to use a new system, called "DOT", for claiming their costs for treating patients. The central principle of DOT is that the amount of the claim is based on the actual care provided, which is only known after the treatment has finished. Previously, the amount of the claim was based on the diagnosis for which average costs would be determined by negotiation between the government, hospitals and insurance companies.

This change confronted the hospitals with a finance management problem. Whereas they previously could claim costs already after diagnosis, they now have to wait until treatment has finished. Since it is unknown ahead of time when treatments will finish, it also unknown when the hospital can expect cash flow for their treatment processes and how large those cash flows will be.

Cash flow prediction for treatment processes seems like a problem that a combination of data and process mining could well help to solve. A treatment process

P. Cudre-Mauroux, P. Ceravolo, and D. Gašević (Eds.): SIMPDA 2012, LNBIP 162, pp. 60–83, 2013.

of one patient, called a *care path*, can be seen as a sequence of activities which determines the so-called *care product* with an associated cost. Based on a data set with all details about completed care paths and associated care products, one could develop a predictor that, given only the start of such a path, could predict the rest of the path and its associated duration and cost.

This paper describes our experiences and results with this case study in data and process mining. The paper specifically addresses a problem with three kinds of noise in our data that significantly affected our results. It is of course common knowledge that noise in the underlying data may affect the result of any kind of data mining including process mining. We encountered, however, three different kinds of noise that were rather specific to our process mining and that were so prevalent that known solutions from literature did not apply. We call these kinds of noise (a) *sequence noise* meaning errors in or uncertainty about the order of events in an event trace, in our case study the order of activities in a care path, (b) *duration noise* meaning noise arising from missing or wrong time stamps for activities and variable duration between activities. (c) *human noise* meaning noise from human errors such as activities in a care path which were the result of a wrong or faulty diagnosis or from a faulty execution of a treatment or procedure.

The reason why sequence uncertainty was so prevalent in our case study is as follows. During the day, a medical specialist typically sees many patients. During a consult or treatment, however, it is often too disruptive for the specialist to update the patient's electronic dossier. It is common practice that (s)he updates the dossiers at the end of day or even later. As a consequence, the modification time of a patient's dossier (which is what is recorded) does not reflect the actual moment of the activity, and it often does not even reflect the order in which the activities took place during that day. Furthermore, it also common that patients undergo several activities and see several specialists on the same day. For example, a hospitalized patient may receive a visit from a specialist doing his/her rounds, receive medication, undergo surgery, results from a blood analysis may be finished, all on the same day. In the case study data sets, we have averages of about 2.5 to 10 activities per day. This inherent noisy timestamp problem and its magnitude causes major *sequence noise* in the underlying data.

Furthermore, the noisy timestamp issue also causes the *duration noise*, as it makes the calculation of the duration between activities very noisy and error prone. Also, much of the *duration noise* comes from the fact that in our case study data two activities are considered by the client to be consecutive even if they are separated by many days or weeks, as long as they are part of the same treatment and no other activities have occurred in between them.

On the other hand, *human noise* arises from human errors which include (i) Noise due to a wrong or faulty diagnosis - this leads to a faulty extra series of process steps at the beginning of the treatment process, (ii) Noise due to faulty execution of the treatment and/or erroneous procedures that need to be repeated. So, our data contains large amounts of *sequence noise* + *human noise*+ *duration noise* and when combined, these make the prospect of noise removal very complicated and difficult to perform. Hence, in our paper we proceed by analysing the noisy data to demonstrate what kind of results one can expect with different data mining analysis techniques.

1.1 Contribution

The contributions of this paper are:

- A case study of applying data and process mining for cash flow prediction in a Dutch hospital.
- Accuracy results for different prediction tasks: given a diagnosis and start of a care path, predict the rest of the path, the duration of the path, the final care product, and the associated cash flow.
- Experimental comparison of the performance of several mining techniques on these tasks.
- Analysis and discussion on the effects of sequence and human noise, both are kinds of data noise specific for process mining which have received little attention in literature.

2 Case Study

Starting January 1st, 2012, Dutch hospitals will use a new system for claiming the costs they make for treating patients. Hospitals claim these costs at patient's insurance companies, who then bill their customers.

2.1 DOT Registration System

The new registration and claiming system addresses the problem that actual costs are unequal to what is charged, in an effort to increase the transparency of cost calculation for provided care. The system is referred to as DOT, which stands for "DBCs towards transparency". A DBC is a combination of diagnosis (D) and treatment (Dutch: behandeling, B).

The central principle of DOT is to decide the care product based on the care provided. Whereas, in the previous registration system the care product was based solely on the diagnosis.

A treatment path in the DOT system consists of the following:

- Care product and associated care product code: This is the name and code given to the treatment performed, which is made up one or more sub paths (Care types).
- Sub path: Is a sequence of activities performed (with one or more Activity types) in a treatment sequence
- Care type and associated care type code: Is the specific name and code associated with the sub path
- Activity type and associated activity type code: Are the activities performed for a particular care type in a sub path

The hospital data is structured into several tables, of which two are important in this case: *Activities* and *Sub paths*. An activity represents an item of work, like surgery or physical therapy, but also, days spent in hospital are considered activities. Groups of activities that are performed to treat the individual patient's diagnosis are called sub paths. They are called sub paths because multiple more-or-less independent sub paths make up the total care path for a patient: the path from the patient registering some complaint to being fully treated. Care paths do not have care products, but sub paths do. Table 1 shows the structure of the activity data, table 2 shows the same for the sub path data.

A care product has an associated cost, which is claimed at a patient's insurance company. To derive which care product a patient has received, the DOT system uses a system called the *grouper*. This grouper consists of rules that specify how care products are derived from performed activities. In the DOT methodology, it is the activities performed that decide the care product. This does not mean that every activity influences the care product, in fact, laboratory work and medicines have no influence on the grouper result. DOTs are processed by the grouper after they have been closed. When a DOT registration is closed depends on the amount of time that has passed since the last activity. If more than the specified number of days has passed, the DOT is marked closed. Different types of activities have different durations after which the DOT is to be marked as closed. The grouper is maintained and operated by the independent DBC Maintenance authority. The rules that make up the grouper are the results of negotiations between the academic Dutch hospitals and insurance companies.

The DOT system should lead to a better matching between actual provided care and associated care product. In turn, this leads to what the insurance companies (and therefore patients) pay.

While a patient is still undergoing treatment, the DOT system poses two problems:

- The hospital does not know how much they will receive for the care they have provided, as they don't know what care product will be associated with the open DOTs.
- The hospital does not know when a DOT is likely to be closed, because a DOT closes only some time after the final treatment. If the patient turns out to require another treatment before the closing date of the DOT (based on the previous treatment), the closing date moves further into the future. Because the hospital does not know when a DOT closes, they also don't know when they can claim the cost of the associated care product.

These problems are in essence process prediction problems. The first involves predicting the process steps, because these process steps dictate the care product. The second problem involves predicting the duration of a process. To see how well different approaches work for this practical case, we test predicting care product, product cost and care duration. For this purpose we use anonymized patient data from a Dutch hospital. The data we have available is based on heart and lung specialties.

Dataset construction. To produce data sets for our experiment, the activity and sub path tables where joined to get a new table with the sub path id, care product, activity code and registration date. A separate table was made for every diagnosis, as the diagnosis is not part of prediction, but is known at the start of a sub path. The produced tables where converted to a *Sub path id − Care product − Activity 1 − . . . − Activity n* format. The data format is explained in detail in section 5.

Because activities are recorded every day and not at the moment they are performed, we do not know the sequence of activities within a day. This leads to *sequence noise* and *duration noise* in the data sets, as the inferred sequence is not necessarily the right one, as we approximate the sequencing by ordering activity codes alphabetically. The *sequence noise*, *duration noise* and the *human noise* increases the complexity of the task of finding the right set of future activities, and then determining its duration.

Note that we did not artificially add noise: the noise is the result of the lack of explicit time-based sequence of activities in the original data, so we had to recover the exact sequence where it was implicit. The approaches we present in this paper for predicting "care products" will have to deal with this fact, as it is a consequence of the way data is stored at the hospital.

3   Problem Formalization

We start with providing notation for and defining the most important concepts in our case study in order to be able to more precisely define the prediction tasks that we distinguish.

3.1   Basic Concepts

An *activity Act* is defined as a label taken from the set of possible DBC codes. A *care path P* is a sequence of activities $P = Act_1, \ldots, Act_n$. We focus our prediction only on subpaths which have a unique care product, so a care path should be interpreted as a subpath. We sometimes denote a care path with $\hat{P}$ to emphasize that it is *closed*, i.e., it belongs to a finished treatment or to a DOT that was closed by the grouper for some reason. The last activity in a closed care path is denoted with Act_{end}; the last activity of an 'incomplete' care path is often denoted with Act_{cur} to emphasize its role as 'current' activity. $P_1 \oplus P_2$ denotes the *concatenation* of P_1 and P_2. $d = d_P$ denotes the *duration* of care path P in terms of time (measured in days in our case study).

The *grouper Grp* is a function that determines the *care product* $C = Grp(\hat{P})$ for a given path $\hat{P}$. The associated *cost* of a care product C is denoted by $cost(C)$.

From a subset of our data, we construct a directed weighted *process graph* $G = (N, E)$, where the nodes N represent activities and the edges e the possibility that one activity can follow another. The *weight of an edge* $w(e)$ is defined as the number of occurrences that these two nodes follow each other in that order. We added a node "Start" to each G to obtain a single starting point for all care paths. We furthermore added the care products as separate terminal nodes, such that for each $\hat{P}$, $Grp(\hat{P}) = C = Act_{end}$.

We assume that the available underlying data is in the form of a set of complete care paths. We chose to work with separate sets of care paths each belonging to one specialty and diagnosis code.

3.2 Prediction Tasks

With the notation above, we can precisely define the prediction tasks we consider in this paper. Let $P = Act_2, \ldots, Act_{cur}$[1], $P' = Act_{cur+1}, \ldots, Act_{end}$, and $\hat{P} = P \oplus P'$. The prediction tasks are:

1a) Given an incomplete care path P, predict the care product C directly (i.e., without considering or predicting P'). We view this prediction task as a *classification* task. We use a subset of the closed care paths belonging to one diagnosis with their associated care product as training data for supervised learning of the classifier.

1b) A predicted care product determines its cost $cost(C)$.

2a) Given a process graph G constructed from a subset of the care paths belonging to one diagnosis, an activity Act_{cur}, and a care product $C = Act_{end}$, predict the path P' in between. We view this prediction task as a *process mining* task. Note that we attempt our prediction given only the current activity Act_{cur} and not the path leading to this activity P. The latter is left to future research.

2b) A predicted path determines its duration, i.e., $d_{P'}$, from which a prediction can be derived for the full duration $d_{\hat{P}}$ given G, Act_{cur}, and C.

In the end the financial department of the hospital is interested in an amount of money (i.e., $cost(Grp(\hat{P}))$) and a moment in time (which can be derived from $d_{\hat{P}}$). It may happen, however, that an entirely wrong care product is predicted, but that it has a similar cost, or that an entirely wrong path P' is predicted with a similar length. In those cases, the prediction of what we are ultimately interested in is close, but rather unjustifiably so. Therefore, we target not only the b-tasks, but also the a-ones. Moreover, examining the results of predicting the rest of the path also provides more insight into the effects of sequence and human noise.

4 Literature Review

4.1 Classifier Algorithms

Prediction tasks 1a and 1b are about assigning a class - the careproduct - to a combination of independent variables - the activities. That is why we consider

---
[1] We start with activity 2, because activity 1 is always "Start".

this to be a classification problem. The field of classifier algorithms divides into four categories: decision tree, clustering, bayesian and neural network classifiers [1]. Besides these classifiers, there are some classifier aggregation techniques that serve to augment the accuracy of individual classifiers.

Decision tree classifiers are based on Hunt's algorithm for growing a tree by selecting attributes from a training set. The attributes are are converted into rules to split the data. Although this can be an accurate technique, the decision tree family of classifiers is sensitive to overfitting or overtraining: training the classifier on a data set with noisy data will include bad (too training-set specific) rules in the tree, reducing accuracy. Algorithms like C4.5 [2] use pruning to prevent overtraining, but the fact remains that decision trees are not effective in very noisy data.

Clustering algorithms for classification use distance (or equivalently: proximity) between instances to group them into clusters or classes. New instances are classified based on their proximity to existing clusters [3]. Examples are nearest-neighbour classifiers and support vector machines.

Bayesian classifiers use Bayes' rule to select the class that has the highest probability given a set of attribute values. Bayes' rule assumes independency of the attributes in a data set, if this is not the case, classifier accuracy could be hindered.

Neural networks are composed of multiple layers of elements that mimic biological neurons, called perceptrons. The perceptrons are trained to give a certain output signal based on some input signal, which is propagated to the next layer of the neural network. The network consists of one input layer, one output layer and multiple hidden layers in between. Neural networks are computationally intensive, but are accurate classifiers. However, neural networks are sensitive to overtraining [3].

Besides techniques consisting of a single (trained) classifier instance, there are techniques that aggregate results from multiple classifiers. Examples are bootstrap aggregating [4], boosting [5] and the Random Forests algorithm [6]. The techniques have the same paradigm: let every individual classifier vote on the class of the data instance, the class with most votes is selected. Breiman found that this principle is especially accurate when the underlying classifier has mixed performance on the dataset [4], something that occurs most often with the less complex algorithms. That is also the case for the Random Forest classifier, which is comprised of multiple randomly grown decision trees. All these aggregation techniques generally display better accuracy than the individual classifiers.

Curram et al. [7] show that decision tree classifiers (such as CART and C4.5 [2]) are outperformed by neural networks. In turn, neural networks are less accurate than the boosting type of classifier aggregation, as shown by Alfaro et al. [8]. Furthermore, the Random Forest classifier [6] has been shown to have good accuracy compared to other classifiers [9] [10]. Therefore, we use boosting (specifically, Freund and Schapire's Adaboost algorithm [5]) and Random

Trace
1. $\{A, B, E, D, G\}$
2. $\{A, C, C\}$
3. $\{B, E, D, F\}$
4. $\{A, C, B, D, G\}$
5. $\{B, D, G\}$
6. $\{A, C, B, E, D, G\}$

(a) Example set of traces

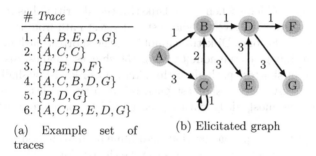

(b) Elicitated graph

Fig. 1. Converting traces to a graph

Forests for care product classification. For some comparison, we also investigate the performance of the CART decision tree algorithm and the naive Bayesian classifier.

4.2   Process Prediction Algorithms

To predict the care activities (path through the process graph), we need for a complete and accurate graph. A possible approach is to take the event log and create a graph that contains every edge in every trace of the log. Consider the set of traces in figure 1a, converted to the graph in figure 1b. Every combination of two subsequent activities is an edge in the graph, every activity is a node. There are no duplicate edges or nodes, but the number of occurrences of two nodes subsequently is noted as the weight of the edge.

The algorithm is naive, as it also includes possibly noisy edges that only occur a few times in the data set. In figure 1b, the $C - C$ edge could be noise, it occurs only once in the data. This makes the naive graph elicitation algorithm inaccurate. To remedy this flaw inherent to the naive approach, the weights of the nodes can be considered as the likelihood of an edge *not* being noise. The more times an edge occurs, the less likely it is to be noise.

Not only individual edges are noisy, also larger sets of edges might be noise. This would result in several nodes connected by low-weight edges. This points to a possible solution: use a path algorithm to determine paths that have the highest weight, or to exclude the paths with the lowest weight. This type of algorithm is known as a shortest-path-algorithm: find a path through the graph that has the lowest possible weight [11].

The most well-known shortest-path-algorithm is Dijkstra's algorithm, which calculates the single shortest path through a series of nodes. This algorithm can be used to determine the distances from a start node to every other node. A distance is the compound weight of the edges towards a node. A modification of Dijkstra's algorithm determines the path that has the highest weight for all its edges combined. To remove (infinite) loops duplicate activities within a trace are relabelled so that they are unique. In figure 1b, this means that C is replaced by C and C', $C - C$ is replaced by $C - C'$.

There are some problems when using Dijkstra: the algorithm has to be run for every node in the graph to give the shortest or longest path from the start node to some other node. Second: Dijkstra cannot give the shortest path between two specific nodes, it only gives the node that has the shortest/highest distance from "start". A different shortest-path algorithm, known as Floyd-Warshall does have this capability: it gives every shortest path between two nodes. This makes it possible to find the most likely path or most unlikely path from one node to another.

Besides shortest path approaches that determine path likelihood by taking the sum of edge weights, another approach is to take the product of the probabilities of the edges. The probability of an edge E from node A to B is determined by the number of times it is traversed divided by the total outgoing weight of A's edges. Given the edge probabilities, the most likely path (based on Bayes' rule, assuming independency of the next edge on previously traversed edges) is the path with the highest product of edge probabilities. This is similar to a Markov chain.

Floyd-Warshall and Dijkstra's algorithm produce the path (set of edges) that has the lowest sum of edge weights. Since both produce the shortest path, both algorithms can be used, but we choose Floyd-Warshall, as it is easier to implement Floyd-Warshall. The adaptation we made on Floyd-Warshall's algorithm does not take the sum of edge weight as the measure of length, but the product of edge weights. We then look for the path (set of edges) that has the maximum product of edge weights: the most likely path.

Finally, we have considered techniques from the process mining field, such as the alpha-algorithm proposed by Van der Aalst *et al.*, the Little Thumb algorithm, InWolve and the suite of algorithms provided by the ProM framework[12, 13, 14, 15, 16, 17, 18, 19, 20, 21, 22]. We have found these algorithms to be unsuitable for our goals, as they do not deal with cycles in the data [23]. Since cycles are a central aspect of the noise in the data available, we do not consider the process mining family of techniques/algorithms useful for our purposes.

4.3   Noise in Process Mining

Datta (1998) [24] defines noise in a business process as an unrelated activity that is not part of any activity of the business process. An example of such an activity would be a phone call or a lunch break [24]. They suggest three techniques for the removal of such noise. The techniques being a stochastic strategy and two algorithmic strategies based on a finite state machine [24].

Weijters and van der Aalst [25] regard noise as (i) a missing activity and/or (ii) randomly swapped activities in the workflow log. To extract the work flow process from the log, Weijters and van der Aalst [25] construct a dependency-frequency (D/F) table (which holds the frequency and order of task co-occurrence or dependencies), from which they construct a D/F graph (the graph of the task dependencies)and finally a work-flow net (or WF net, that represents the D/F graph along with the splits and joins) using heuristics [25].

Both Agrawal et al. [26] as well as Huang and Yang's [27] define noise as instances when executed activities were not collocated, timestamps of the activities are mistakenly recorded or exceptions that deviate from the normal processing order [27]. They furthermore use similar probabilistic reasoning and estimate how much error their process discovery algorithm would give due to the noise [27].

As we have explained earlier, the noise that we describe in our paper is however in a slightly different league from the noise described in process literature. Hence instead of cleaning the noise, we take a different approach and show what results one can expect by analysing very noisy data with different data mining techniques.

5   Design of Case Analysis

Table 4 shows the main attributes of the data in the three data sets we use. Each set contains of rows of activities that are each represented by a code. Each specific activity, such as "Perform surgery X" has its own unique code in the DOT method. Only when creating a graph are these activities relabelled to remove loops, so table 4 does not include relabelled activities.

An idea of the extent of noise in the data is given by the *Average number of activities per day* and *Average path length* in table 4: large parts of the sequence of activities are uncertain and may well be wrong - although we have no way of knowing. Hence, in order to reduce the *sequence noise* a subset of the data was used for the classification tasks. To create the subset, we removed the activity codes for lab-work and medications from the raw data, because these were known to have no effect on the grouper result of a sub path. The characteristics of the data set for the classification task are shown in table 5. The effect of removing these activities is best shown by the shorter average path length and the fewer average number of activities per day.

We created three data sets from the hospital's cardiology and lung care department data. Each set contained the data from patients with a specific diagnosis. The three diagnoses were pericarditis (an inflammation of the heart), represented by code 320.701; angina pectoris (chest pain), diagnosis code 320.202; and malignant lung cancer, diagnosis code 302.701. We chose these data sets as they were clearly quite different, not only in size, but also in the amount of sequence noise, path length and duration. This made the sets well suited to test the prediction approaches presented in this paper on a broad sample of actual process data. Also, only data from the heart and lung specialties was available to us for this research.

Table 3 shows the columns that make up each dataset, as well as some example content. The number of columns varies between 286 + 2 for the dataset 302.66 to 703+2 for 302.701.

Table 1. The performed activity table

Activity id	Care path id	Sub path id	Care activity code
The unique id of the performed activity	Unique identifier of the care path containing the sub path the activity is part of	The identifier of the sub path the activity is part of	A (non-unique) code describing the activity

(cont.d)	Date	Performing specialty	Amount
	The date on which the activity was registered	The specialty performing the activity	The quantity of the activity, for example used for medication

Table 2. The sub path table

Sub path id	Care path id	Start date	End date	Medical specialty
The unique identifier of the sub path	The care path the sub path is part of	The date on which the sub path was opened	The date on which the sub path was closed	The specialty responsible for this sub path

(cont.d)	Care type	Diagnosis code	Care product code	Closure reason
	The care type of the sub path.	A code for the diagnosis that led to this sub path	The care product for the closed sub path	The reason the care path was closed, for example, enough time had expired.

Table 3. The columns in each dataset and example dataset contents

Columns:

Sub path	Care product	Activity 0	Activity 1	Activity 2	...	Activity n

Example contents:

Sub path	Care product	Activity 0	Activity 1	Activity 2		Activity n
1700750	979001104	33229	299999	33231	190205	33229
40176052	99499015	33285	10320			

Table 4. Main characteristics of the data sets

	320.202	320.701	302.66
Number of sub paths	9950	274	1985
Number of unique activity codes	315	190	239
Number of unique care products	22	6	11
Largest path length $|\hat{P}|$	559	703	286
Average path length $|\hat{P}|$	15.99	41.36	15.07
Average number of activities per day	6.79	10.13	2.49
Average duration (in days)	35.81	31.23	126.22

Table 5. Main characteristics of the classification data set. Lab-activities are excluded from this set.

	320.202	320.701	302.66
Number of sub paths	9950	274	1985
Number of unique activity codes	160	70	149
Number of unique care products	22	6	11
Largest path length $|\hat{P}|$	152	174	104
Average path length $|\hat{P}|$	6.91	11.35	11.39
Average number of activities per day	3.18	3.03	2.15

Using these data sets we performed the prediction tasks in section 3.2. The setup per task is:

Task 1a) *Predict care product C from P.* We wanted to know the prediction accuracy for different lengths of incomplete care path C. So, for path length $1 \ldots 30$, we took a sample of the closed care paths of at least that length for training - the remainder was the test set. We took the first n activities per path P in the training and test sets. The training set of paths of length n in combination with the care product C were then used to train a classifier. The test set of paths was used to assess the performance of the classifier.

Task 1b) *Predict care product cost:* $cost(Grp(\hat{P}))$. This task builds on task 1a: The same procedure was used to determine a care product C, but now we looked up the cost $cost(C)$. The predicted cost for a partial path P was then compared to the cost of the actual care product $C = Grp(\hat{P})$. We measured the difference in predicted cost and actual cost E_{total} as well as the average absolute error $E_{average}$:

$$E_{total} = \frac{|\sum cost(Grp(\hat{P})), \sum cost(Grp(\hat{P}'))|}{\sum cost(Grp(\hat{P}))}$$

$$E_{average} = \frac{|cost(Grp(\hat{P})), cost(Grp(\hat{P}'))|}{|\hat{P}|}$$

Task 2a) *Predict P'..* Here, we determined the precision and recall of predicting the path P' between Act_{cur} and Act_{end}. So, we created the process graph G from a training sample of the complete paths $\hat{P}$. The remainder of the data set was then used for testing. Next, for activity $Act_n \in \hat{P}, n \in \{1 \ldots 15\}$ for each complete path in the test set we predicted P'. We compared P' to the sub path $Act_n \ldots Act_{cur} \in \hat{P}$ and determined the precision and recall of the prediction.

In order to make sense of the results of both algorithms, we introduce two performance metrics: precision and recall. These metrics are based on finding the number of true positives, false positives and false negatives. True positives are the number of activities that occur in both predicted P' and actual P'. False positives are the activities that are predicted but do not occur in the actual path. False negatives are the number of nodes that are not in the predicted P', but are in the actual P'.

Given these three numbers, precision and recall are calculated by:

$$Precision = \frac{tp}{tp + fp} \qquad Recall = \frac{tp}{tp + fn}$$

Task 2b) *Predict $d_{\hat{P}}$.* Given the predicted path P' we created a predicted complete path $\hat{P}' = P \oplus P'$. We looked up the duration for each edge $e \in \hat{P}'$, where duration is the average time difference between $Act_a, Act_b \in e$ for all occurrences of e in the training set. These durations were combined into a duration prediction $d_{\hat{P}'}$ and were compared to the actual time difference between $Act_1, Act_{end} \in \hat{P}$.

6   Results

6.1   Task 1a: Predicting $Grp(C)$

Table 6 shows the results for the care product predicting task. As expected from the literature review, we found the Random Forest algorithm to be the best performing classifier in terms of accuracy. This table shows the accuracy for care product prediction for path lengths 1 to 30. Table 7 shows the Random Forest results when only taking paths that have at least length 10. Figures 3 and 4 show the accompanying Random Forest accuracy plots for these two tables. The plot in figure 2 shows the number of paths of length n for each of the data sets, which also explains the cut off point of 10 activities: the number of complete paths $\hat{P}$ with more than 10 activities is low.

To put the Random Forest accuracy in perspective, we compare the results to a more naive approach: take a random guess of care product. This would result a chance of predicting the care product of 16 percent maximum, given that the data set with the fewest number of care products has six products. The Random Forest classifier seems to outperform this method. The comparison with a somewhat less naive approach, taking the most occurring care product is shown in figure 3 and 4. This shows that the Random Forest classifier is always better than a *random classifier* for two of the three datasets. The dashed lines in both plots represent the expected accuracy when always selecting the most occurring care product.

The accuracy of the classifier does not improve with a larger path length $|P|$. This is the result of noise: as the number of activities in sequence increases, the likelihood of that sequence containing noise also increases. This noise has a negative effect on the classifier's performance.

Table 6. Accuracy (fraction correct of total) of prediction of care product for path lengths $|P| = 1 \ldots n$

Training set: 50%.

320.202

	Min	Max	Mean	Median
Random Forest	0.30	0.69	0.47	0.44
CART Tree	0.24	0.49	0.37	0.35
Adaboost.M1	0.24	0.59	0.41	0.40
Naive Bayes	0.21	0.45	0.30	0.28
k Nearest Neighbour	0.20	0.53	0.31	0.28

320.701

	Min	Max	Mean	Median
Random Forest	0.44	0.71	0.55	0.53
CART Tree	0.00	0.75	0.47	0.50
Adaboost.M1	0.11	0.75	0.52	0.56
Naive Bayes	0.00	0.80	0.45	0.45
k Nearest Neighbour	0.11	0.63	0.42	0.43

302.66

	Min	Max	Mean	Median
Random Forest	0.28	0.55	0.40	0.38
CART Tree	0.16	0.55	0.35	0.33
Adaboost.M1	0.24	0.56	0.38	0.35
Naive Bayes	0.19	0.46	0.31	0.29
k Nearest Neighbour	0.20	0.45	0.28	0.25

6.2 Task 1b: Predicting $cost(Grp(C))$

The goal of predicting a care product C is predicting the cost, so the hospital knows what it will receive for provided care. Table 8 shows the results for a test of predicting $cost(C)$ in terms of the total error fraction E_{total}. Figure 5 shows the average absolute error $E_{average}$ in predicting the cost. These errors cancel each other out in some cases, which explains the lower best performance

Table 7. Prediction of care product from path length $|P| = 1 \ldots 10$

	Min	Max	Mean	Median
320.202	0.43	0.58	0.49	0.50
320.701	0.36	0.54	0.43	0.40
302.66	0.43	0.48	0.45	0.45

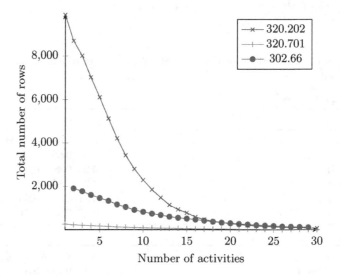

Fig. 2. Number of sub paths of length $|P| \geq 1 \ldots n$

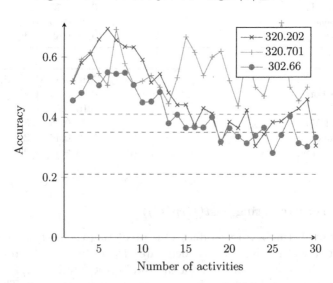

Fig. 3. Random Forest accuracy for path length $|P| = 1 \ldots n$. Dashed lines represent the expected accuracy when randomly selecting care product

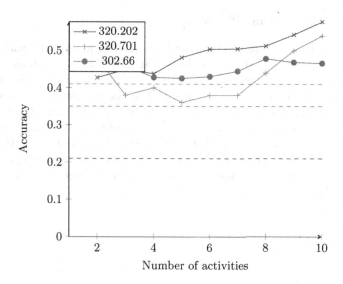

Fig. 4. Random Forest accuracy for path length $|P| = 1 \ldots 10$ and $|\hat{P}| \geq 10$. Dashed lines represent the expected accuracy when randomly selecting care product

results shown in table 8 than in 5. The test shows again that Random Forests provide decent predictions of the care product: Even though it averages around 40 percent correctly predicted care products, the 60 percent it predicts wrong apparently has a similar cost.

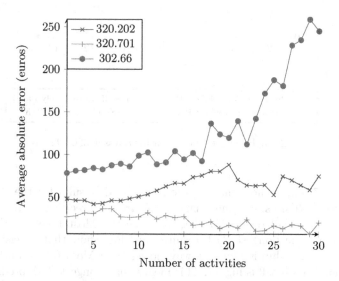

Fig. 5. Random Forest cost error for path length $|P| = 1 \ldots n$

6.3   Task 2: Predicting P' and $d_{\hat{p}}$

Path search algorithms are the tools we have used to predict the path P'. The algorithms we use are variations on Floyd-Warshalls shortest path algorithm, as discussed above. We have experimented with two algorithms, that both take an activity and a care product and attempt to find the activities in between.

The first variation, called Longest Path, returns P' such that the sum of the weight of the edges is maximized. Here, the edge weight is the number of times that the start and end node of that edge occurred consecutively. The second variation, called Most Likely Path, returns P' such that the product of edge weights is maximized. In this scenario, the weight of an edge shows the likeliness of that edge being traversed.

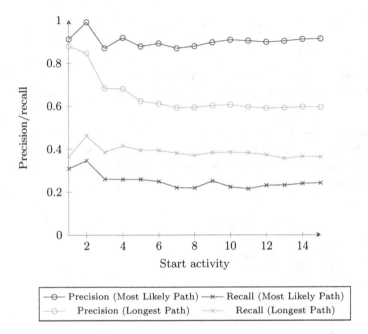

Fig. 6. Precision and recall for dataset 320.202

Figures 6, 7 and 8 show the precision and recall results for the three data sets. In all these figures, the lines marked "o" are the precision results and lines marked "x" are recall results. Darker lines are the Most Likely Path results, lighter lines are the Longest Path result. The plots show that precision of the Longest Path algorithm is almost always lower than Most Likely Path, but on the other hand, its recall is higher. This means that Longest Path predicts more of the actual activities, but also predicts activities that do not actually occur.

Tables 9a and 9b show the effect of the path predictions on the duration prediction for Most Likely Path and Longest Path, respectively. Both algorithms

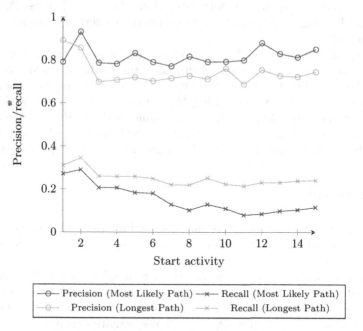

Fig. 7. Precision and recall for dataset 320.701

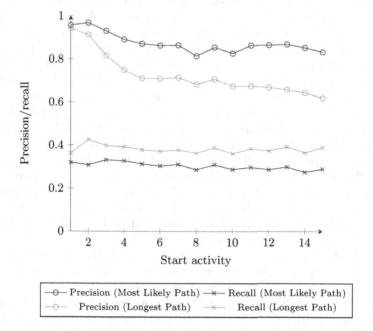

Fig. 8. Precision and recall for dataset 302.66

show lower predictions than the actual durations. Again, the Longest Path shows the effect of having greater recall: the predicted path lengths are longer and closer to the actual.

The average duration prediction error in table 9 seems reasonable at up to around two weeks. This changes when taking the predicted durations into account: they are close to zero for the Most Likely Path. Of course, this problem is related to the lack of recall using MLP. Longest Path is performing a bit better, but still gives short predicted durations. If the predicted path is much shorter than the actual the predicted duration will also be much lower: duration is calculated as the sum of average edge durations in the predicted path.

7   Discussion

We can analyse and discuss the results obtained above in the light of the three different kinds of noise in the data: namely, the *sequence noise, duration noise* and *human noise*.

Let us consider the results from each of the goals of this paper. Beginning with (1a), predicting $Grp(C)$. The accuracy of the prediction does not go up when we consider sequences containing a greater number of activities 3, the reason for this is that we assumed a sequence for the activities conducted every day - *sequence noise*. Hence, when we have a greater number of activities on a certain day, we would expect a larger *sequence noise*, and as a result the accuracy of predicting the care product would decrease. For this measurement we used a subset of the actual dataset, the main characteristics of which can be seen in Table 5. Comparing the average number of activities in Table 5, with those in Table 4, we can clearly see that there has been a large decrease in *sequence noise*, especially for data sets *320.202* and *320.701*. Hence, we can say that for this task the noise in the data sets is nearly of the same magnitude. This is the case as the *duration noise* and the *human noise* (that does not influence the *sequence noise*) do not affect the classification task. So, it comes as no surprise that the results of the classification (taking the mean Random Forest values) shown in Table 6, are nearly the same for the three sets of data. Added to this is the fact that if we use the best performing Random Forest classifier, it helps in taking care of some of the error in the data. Furthermore, if we consider only subpaths whose path lengths are at least 10, we see from Fig.4 that the accuracy for predicting the care product has the same trend for all the three data sets. However, we do notice from Figure 3 (and even from Figure 4) that the accuracy of predicting the path for data set *302.66* drops after about 15 activities and gets even worse than randomly selecting the care product. One could attribute this to a certain amount of *human noise* present in the data set *302.66*, which could be greater than the *human noise* present in the other two data sets.

Task (1b), predicting $cost(Grp(C))$, again we see the Random Forest algorithm performing the best. Though the mean values of E_{total} is similar for *320.202* and *320.701*, it is higher for the data set *302.66*. This maybe explained partly due to the greater amount of *human noise* present in the data set *302.66*,

Table 8. E_{total} in prediction of $cost(C)$ for path lengths $|P| = 1\ldots 30$

Training set: 50%.

320.202

	Best	Worst	Mean	Median
Random Forest	0.00	0.15	0.07	0.06
CART Tree	0.00	0.21	0.07	0.07
Adaboost.M1	0.00	0.20	0.08	0.07
Naive Bayes	0.16	0.62	0.33	0.25
k Nearest Neighbour	0.00	0.21	0.06	0.03

320.701

	Best	Worst	Mean	Median
Random Forest	0.00	0.18	0.06	0.04
CART Tree	0.00	0.23	0.07	0.05
Adaboost.M1	0.01	0.20	0.07	0.05
Naive Bayes	0.01	0.43	0.15	0.11
k Nearest Neighbour	0.01	0.16	0.07	0.07

302.66

	Best	Worst	Mean	Median
Random Forest	0.00	0.51	0.17	0.14
CART Tree	0.00	0.47	0.15	0.14
Adaboost.M1	0.00	0.51	0.18	0.18
Naive Bayes	0.01	0.72	0.39	0.52
k Nearest Neighbour	0.00	0.24	0.11	0.10

Table 9. Average predicted duration and prediction error, in days

(a) Most Likely Path				(b) Longest Path			
	202	701	66		202	701	66
Average predicted duration	0.4	0.7	0.7	Average predicted duration	3.7	0.5	9.5
Average prediction error(days)	11.7	3.6	14.3	Average prediction error	8.4	3.7	5.5
Average % of prediction error	33%	11.5%	11.3%	Average % of prediction error	24%	11.8%	4.1%

which also explains the trend of E_{total} in Figure 5, where similar to Figure 3 the average absolute error of *302.66* increases as compared to the other two data sets.

Task (2a), predicting P'. For tasks (2a) and (2b) we need to keep in mind that we use the actual raw data set (Table 4) and not a subset(Table 5) as in the case of Tasks (1a) and (1b), the reason being that the laboratory tests do affect the result of the tasks (2a) and (2b). From Figure 6, Figure 7 and Figure 8 we see that recall is always lower than the precision. The reason behind this is that the methods we used to predict the path, returned paths that were shorter than the actual paths the data had and hence the recall is low. While on the other hand, the precision is higher as the predicted activities are mostly in the actual path. We also see a consistently high trend of precision for the Most Likely Path algorithm for all three data sets, which, as expected, has a decreasing trend for *302.66*. This could be due to the higher presence of *human noise* (as we noted earlier). However, when we observe the trend for the Recall values (in Figure 6, Figure 7 and Figure 8) we see that the performance is in the order: Recall(*302.66*) > Recall(*320.202*) > Recall(*320.701*). We think this is an indicator of the *sequence noise* present in the data sets which also follows the same trend, going by the number of average activities per day in Table 4. We think that *sequence noise* affects the recall more than the precision as the greater the *sequence noise* the greater is the number of actual activities not in a particular path, i.e. the false negatives are higher.

Task (2b), predicting $d_{\hat{p}}$. From Table 9a and Table 9a, we see that the Longest Path gives less error than the Most Likely Path algorithm. This could be as we explained earlier a consequence of the Longest Path algorithm having a higher recall (Figure 6, Figure 7 and Figure 8) and hence having fewer false negatives and being closer to the original path. We also notice that the prediction error is higher for *320.202* compared to *302.66* and *320.701*. This result seemingly contradicts the trend of *sequence noise* between the data sets *320.202* and *302.66*, as discussed in the earlier paragraph. We can however reason this result, by arguing that *sequence noise* does not affect the duration prediction as much as *duration noise* does. For example, in our Hospital data two identical subpaths (with the same activities and in the same order) can have entirely different durations, because of the different timestamps and the different intervals between activities (that can vary from a couple of minutes to weeks. A practical example of such a case is when an young and a relatively old patient is admitted in a Hospital for a similar fracture or injury. Both the individuals would get a similar treatment(Care product - and would have the same subpaths, however, the older patient could take a much longer time to recover. Hence, we can reason the data in Table 9a and Table 9a by concluding that the data set *320.202* has greater *duration noise* than the data sets *302.66* and *320.701*.

An issue prevalent in the data sets and which could be one of the major causes of the *duration noise* and *human noise*, is the fact that all patients, irrespective of their particular demographic background, are subjected to the same treatment procedures. This would cause a lot of *human noise*, apart from the *duration noise* (as explained in the earlier paragraph), as we can expect that all patients would not respond to the same treatments in an identical manner. Hence, the likelihood of faulty treatment procedures (*human noise*) increases.

8   Conclusions

In our case study we have shown how one can combine data and process mining techniques for forecasting cash flow in Dutch hospitals. The techniques and processes we have demonstrated, we think can be used to analyse the cash flow for other hospitals with a few modifications.

For patients still undergoing treatment, we managed to, given a diagnosis and past activities, predict the care product with an accuracy of a little less than 50% (for all the 3 sets of data), which was still (on an average) better than randomly predicting the care product. On the other hand, for predicting the associated cost of the Care, we obtained an error of of less than 10% for 2 data sets and 17% for one. These results were obtained with a Random Forest classifier which was shown to perform the best. This shows that our method of predicting is a viable solution to the cash flow problem we raised in the introduction.

We furthermore managed to, given a diagnosis, the most recent activity, and the (predicted) care product, predict the remaining activities using two algorithms (Longest Path and Most Likely Path). We achieved a precision of about 80% for the Most Likely Path and about 60-70% for the Longest Path algorithm and a recall of 35-40% (for the three data sets). We algorithms predicted the associated duration with an error of about 30% for 202 and around 11% for the data sets 701 and 66.

The case study is of particular interest because of the causes for the modest to weak prediction results for the process mining tasks. In our opinion, three process mining-specific types of noise significantly affected our results: sequence noise, human noise, and time noise. In this paper, we thoroughly discuss the causes for these types of noise and their effects.

From the predicted cost and duration, it is possible to derive a prediction (albeit with some error) for the future cash flow for all patients currently undergoing treatment. This we think is a major contribution to research and practice, as given the noise in the data, we were still able to make reasonable predictions which could cheaper and more practical than asking a few expert practitioners.

Future work can include attempts at removing or decreasing the different kinds of noise in the data and then carrying out the prediction analysis to compare the results.

References

[1] Han, J., Kamber, M., Pei, J.: Data mining: concepts and techniques. Morgan Kaufmann Pub. (2011)
[2] Quinlan, J.: Improved use of continuous attributes. Journal of Artificial Intelligence Research 4 (1996)
[3] Alpaydın, E.: Introduction to Machine Learning. The MIT Press (2004)
[4] Breiman, L.: Bagging predictors. Machine Learning 24, 123–140 (1996)
[5] Freund, Y., Schapire, R.: Experiments with a new boosting algorithm. In: Machine Learning: Proceedings of the Thirteenth International Conference, pp. 148–156 (1996)
[6] Breiman, L.: Random forests. Machine Learning 45(1), 5–32 (2001)

[7] Curram, S., Mingers, J.: Neural networks, decision tree induction and discriminant analysis: an empirical comparison. Journal of the Operational Research Society, 440–450 (1994)

[8] Alfaro, E., García, N., Gámez, M., Elizondo, D.: Bankruptcy forecasting: An empirical comparison of adaboost and neural networks. Decision Support Systems 45(1), 110–122 (2008)

[9] Banfield, R., Hall, L., Bowyer, K., Kegelmeyer, W.: A comparison of decision tree ensemble creation techniques. IEEE Transactions on Pattern Analysis and Machine Intelligence 29(1), 173–180 (2007)

[10] Gislason, P., Benediktsson, J., Sveinsson, J.: Random forests for land cover classification. Pattern Recognition Letters 27(4), 294–300 (2006)

[11] Baase, S., Gelder, A.: Computer algorithms: introduction to design and analysis. Addison-Wesley (2000)

[12] van der Aalst, W.M.P., van Dongen, B., Herbst, J., Maruster, L., Schimm, G., Weijters, A.J.M.M.: Workflow mining: a survey of issues and approaches. Data & Knowledge Engineering 47(2), 237–267 (2003)

[13] van der Aalst, W.M.P., Weijters, A.J.M.M., Maruster, L.: Workflow mining: Discovering process models from event logs. IEEE Transactions on Knowledge and Data Engineering 16(9), 1128–1143 (2004)

[14] van der Aalst, W.M.P., Günther, C.: Finding structure in unstructured processes: The case for process mining. In: Seventh International Conference on Application of Concurrency to System Design, ACSD 2007, pp. 3–12. IEEE (2007)

[15] van der Aalst, W.M.P., Reijers, H., Weijters, A.J.M.M., van Dongen, B., de Medeiros, A.K.A., Song, M., Verbeek, H.: Business process mining: An industrial application. Information Systems 32, 713–732 (2007)

[16] van der Aalst, W.M.P., Rubin, V., Verbeek, H.M.W., van Dongen, B., Kindler, E., Günther, C.: Process mining: a two-step approach to balance between underfitting and overfitting. Software and Systems Modeling 9(1), 87–111 (2010)

[17] van der Aalst, W.M.P., Weijters, A.J.M.M.: Process mining. Process-Aware Information Systems, 235–255 (2011)

[18] Mans, R., Schonenberg, M., Song, M., van der Aalst, W.M.P., Bakker, P.: Application of process mining in healthcare–a case study in a dutch hospital. Biomedical Engineering Systems and Technologies, 425–438 (2009)

[19] Günther, C., Rinderle-Ma, S., Reichert, M., Van Der Aalst, W.M.P.: Using process mining to learn from process changes in evolutionary systems. International Journal of Business Process Integration and Management 3(1), 61–78 (2008)

[20] Weijters, A.J.M.M., van der Aalst, W.M.P.: Rediscovering workflow models from event-based data using little thumb. Integrated Computer Aided Engineering 10(2), 151–162 (2003)

[21] Herbst, J., Karagiannis, D.: Workflow mining with InWoLve. Computers in Industry 53, 245–264 (2004)

[22] van Dongen, B.F., de Medeiros, A.K.A., Verbeek, H.M.W., Weijters, A.J.M.M., van der Aalst, W.M.P.: The prom framework: A new era in process mining tool support. In: Ciardo, G., Darondeau, P. (eds.) ICATPN 2005. LNCS, vol. 3536, pp. 444–454. Springer, Heidelberg (2005)

[23] Kiepuszewski, B., ter Hofstede, A., van der Aalst, W.M.P.: Fundamentals of control flow in workflows. Acta Informatica 39, 143–209 (2003)

[24] Datta, A.: Automating the discovery of as-is business process models: Probabilistic and algorithmic approaches. Information Systems Research 9(3), 275–301 (1998)

[25] Weijters, A.J.M.M., van der Aalst, W.M.P.: Rediscovering workflow models from event-based data using little thumb. Integrated Computer Aided Engineering 10, 151–162 (2003)

[26] Agrawal, R., Gunopulos, D., Leymann, F.: Mining process models from workflow logs. In: Schek, H.-J., Saltor, F., Ramos, I., Alonso, G. (eds.) EDBT 1998. LNCS, vol. 1377, pp. 469–483. Springer, Heidelberg (1998)

[27] Hwang, S., Yang, W.: On the discovery of process models from their instances. Decision Support Systems 34(1), 41–57 (2002)

Towards Automatic Capturing of Semi-structured Process Provenance

Andreas Wombacher and Mohammad Rezwanul Huq

University of Twente, 7500 AE Enschede, The Netherlands
{a.wombacher,m.r.huq}@utwente.nl

Abstract. Often data processing is not implemented by a workflow system or an integration application but is performed manually by humans along the lines of a more or less specified procedure. Collecting provenance information in semi-structured processes can not be automated. Further, manual collection of provenance information is error prone and time consuming. Therefore, we propose to infer provenance information based on the file read and write access of users. The derived provenance information is complete, but has a low precision. Therefore, we propose further to introducing organizational guidelines in order to improve the precision of the inferred provenance information.

1 Introduction

Semi-structured processes are business or scientific workflows, where the execution of the workflow is not completely controlled by a workflow engine, i.e., an implementation of a formal workflow model. Examples can be found in scientific communities where a workflow of data cleansing, data analysis and data publishing is informally described but the workflow is not automated. Other examples can be found in scenarios where several people potentially from different organizations cooperate e.g. in creating a yearly progress report or writing a scientific paper.

The lack of a workflow system controlling the process execution and the usage of non-provenance aware applications for executing activities, i.e., the application is not recording provenance information itself, makes it difficult to answer provenance questions like 'Who did what when and how?'. Since the provenance question can not be answered it is hard to assess the quality of the output of the semi-structured process, like e.g. the quality of data in a report. In many scenarios, manual provenance acquisition is infeasible since it is too time consuming, and therefore too costly. Thus, the aim is to explore automatic provenance capturing in semi-structured processes.

By analyzing the execution of semi-structured processes the following observations can be made: First, the result of an activity is a single document or a set of documents.

Second, information contributing to a document is often copied from previous activity results, i.e., other documents. Thus, for an activity the source document must have been read/opened before the information can be copied and saved in the target document. Therefore, each document which has been read before a save operation is performed may have contributed to the target document.

P. Cudre-Mauroux, P. Ceravolo, and D. Gašević (Eds.): SIMPDA 2012, LNBIP 162, pp. 84–99, 2013.

Third, users of ICT systems organize their information often in hierarchical structures like e.g. directories. Thus, for a certain activity a user has to perform specific directories are relevant.

Fourth, several revisions of documents are issued over time, where the revision number may or may not be explicated in the file name. However, the process handles potentially various revisions of a document.

In this paper, an approach to automatically capture provenance information for semi-structured processes is proposed. The basic idea is that all documents, which have been read by a user, may have contributed to a document saved later on. Since this derived provenance is rather imprecise it is proposed to facilitate knowledge of the process, of the revisions of documents, and of the organization of the documents by the user to increase the precision. In particular, it is proposed to introduce organizational directives, i.e., guidelines for the user on how to organize information relevant to the data processing workflow. The more strict these guidelines are the higher the precision that can be achieved. However, strict guidelines lower the degree of freedom for a user to organize 'her' data and therefore may result in non complying users effecting the data quality.

In a next step we investigate the mining of provenance relations from the collected version information of the used files without considering the organizational directives. It turns out that actions performed by a script or program can be mined well while manual modifications of files can be detected but the provenance can not be mined. The results are qualitatively compared with the provenance information derived from the organizational directives.

The proposed approach is based on a WebDAV infrastructure which supports version control of files. The proposed approach has been implemented and evaluated on the paper writing of this paper.

In the following related work is discussed (Sect 2) before a use case is introduced (Sect 3). The approach is presented on a conceptual level (Sect 4) while a more technical view on the derivation of provenance information is provided in Sect 5. Next an evaluation of the proposed approach (Sect 6) is presented followed by a description of the mining approach (Sect 7) and the conclusions (Sect 8).

2   Related Work

Automatic collection of provenance information is often applied in e-science workflow systems, like e.g. Kepler [1] or Taverna [2]. Most systems even rely on exchanging data via files. In previous work [3] we investigated inference of provenance information for stream processing workflow system using a temporal model. However, the workflow system is executing a workflow and all involved tasks are executed automatically which is a major difference to our requirements.

Provenance information is also collected in closed systems like e.g. a data warehouse [4] or a relational database [5]. The level of granularity in these approaches varies between fine-grained and coarse-grained data provenance [6]. The fact that the provenance acquisition is limited to the system makes it infeasible for our scenario.

Automatic collection of provenance information focusing on the exchange of information is addressed in various ways. In [7] provenance information is captured by

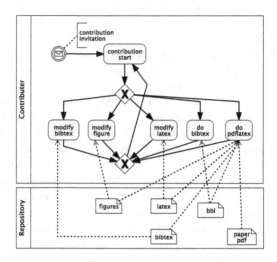

Fig. 1. Editing phase of a paper writing process

monitoring a service bus where the invoked services must not be provenance aware. Although this is quite close to the scenario at hand, in this case all processing has to be implemented as a service, which is not the case in our case. The approach in [8] records file manipulation operations including system variables and their changes. However, this approach is limited to a single computer since it is woven into the kernel of a linux system which differs significantly from our scenario. This may be the closest match to our approach. In [9] the authors automatically collect provenance information based on events recorded by browsers. However, the approach is limited to a single application.

Systems for storing provenance information are e.g. Tupelo2 [10] or Karma2 [11]. These systems provide an into store provenance information and provide means to query the data. The provenance information derived by the proposed approach could be stored in such a system. Further, the acquired provenance information could be made accessible in different provenance models like e.g. the Open Provenance Model (OPM) [12] or the value centric model (TVC) [13]. However these are just alternative representations of the derived provenance data, while the focus of this paper is on acquiring the provenance data rather then how to represent them.

3   Use Case

The process used as a running example is writing this technical paper with multiple authors in Latex. After initializing the project, two authors are writing together on the same paper. A BPMN notation of the process is depicted in Fig 1[1].

---

[1] Created with http://oryx-project.org/

In the process the following activities can be executed in arbitrary order with an arbitrary number of repetitions:

- creation, update and conversion activities on figures and graphics files (modify figure)
- creation and update activities on bibtex files (modify bibtex)
- creation and update activities on latex files (modify latex)
- pdflatex activity, for creating a pdf file of the paper (do pdflatex)
- bibtex activity, for creating the bibliography file (bbl file) related to the paper (do bibtex)

All activities result in a single result or output document/file. There are two authors involved in this particular paper writing process, which makes it interesting to determine whether a specific revision of a generated pdf file contains all the latest file revisions. Especially whether all figures have been properly converted before executing the pdflatex task. Further, it can be inferred whether in a specific pdf revision of the paper all indexes and the bibliographic information is up to date, since this requires the following task sequence: pdflatex - bibtex - pdflatex -pdflatex.

4 Approach

The aim is to infer provenance information from a semi-structured process execution without the user providing any information and the legacy applications not being provenance aware. Since files are used to exchange information between different activities of the process, the approach is based on documenting data manipulations on files. Combining file manipulation information with data dependencies of the process allows to infer which revision of which file may have contributed to a revision of a file written by a particular user.

The intuition is that data processing is based on zero or several input files producing one output file. In particular, all files which have been opened before the point in time a file is saved potentially contributed to the creation of the saved file. These derived provenance relations have a very low precision, i.e., are too broad especially when considering the amount of files opened in the coarse of a day at a desktop computer.

Therefore, we propose to facilitate knowledge of the process to derive organizational directives, i.e., rules for the user performing the data processing activities to be able to associate files with specific activities in the process. Thus, they support inference of provenance information. Organizational directives give a responsibility to the user without technically enforcing the directive. Organizational directives are widely implemented in organizations, like e.g. you are not allowed to install software on your company laptop, you are not allowed to download copyright protected material, you are obliged to make backups or to encrypt your hard disc.

The approach is depicted in Fig 2. Based on the semi-structured process (lower left corner of Fig 2) the organizational directives are derived, of which inference rules for provenance relations can be derived. Further, the derived provenance inference rules are continuously applied on the automatically acquired file access information (gray box in Fig 2) to derive provenance relations. In particular, a relation exists, if a file

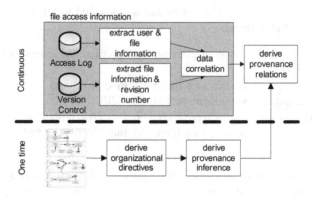

Fig. 2. Prototype data processing

with revision v_s has been written after a file with revision v_r has been read and $v_r < v_s$ assuming a global version scheme. In the following the basic derivation of provenance relations (continuous part) and the creation of organizational directives (one time part) of the approach are discussed in more detail after the infrastructure and some generic directives are introduced. The inferred provenance information is made accessible via a web application [2].

4.1   Infrastructure to Capture User Behavior

The proposed approach is based on a Web based infrastructure for file storage and version control, in particular, a WebDAV infrastructure facilitating a Subversion (SVN) as version control system. WebDAV is an HTTP based protocol for managing content on a web server, i.e., accessing, writing and moving files. WebDAV seamlessly integrates into various operating systems, i.e., the WebDAV server appears as a mounted network drive and therefore is intuitively usable also for basic ICT users.

Further, since the WebDAV protocol is based on HTTP it easily supports distributed and cross-organizational scenarios, like the motivating example in Sect 3. Furthermore, WebDAV can also be used for information dissemination. From a provenance capturing point of view, WebDAV has the advantage that it supports logging of file read operations via the web server access log and logging of write operations via the underlying version control system. In particular, the web server access log documents the read (GET), write (PUT), move (MOVE), and delete (DELETE) operations applied to files on the server. Further, the Subversion (SVN) version control log documents write (Add, Modify), delete (Delete) and move (combined Add and Delete) operations. Both sources (see content gray box in Fig 2) are required to capture the file information and therefore have to be correlated (see Sect 5.1).

The choice of using WebDAV seems a bit outdated compared to currently existing offers like DropBox, Google Drive or Microsoft SkyDrive. While all these offers provide history information comparable to the SVN functionality, the size of the history information maintained is limited. The history information documents the write access

---

[2] Accessible at http://www.sensordatalab.org/offline_provenance_web/

to files, however, the read access to files is not documented there. Furthermore, the products synchronize the files with all related devices instantly, thus, every read operation by another person after a change is a local file operation, which is not documented. Therefore, WebDAV although old fashioned allows to track read access since the WebDAV functionality is under our own control.

4.2 Generic Organizational Directives

The proposed approach is based on observing the handling of files. Thus, some generic directives are needed to ensure that the handling of files can be observed at the first place. Similar to directives in organizations that all important information has to be saved on a network drive because local disks are not backuped, a data security directive is introduced requiring the user to save data on a mounted network drive, i.e., a WebDAV server.

Directive 1 (Data Security). *Users must save all files related to a semi-structured process on the network drive. Thus, it is not allowed to store files related to the process on a local disc.*

This directive is necessary, since local file systems can not be monitored with regard to file handling that easily from outside the computer.

A standard directive in organizations is that login and password information must not be shared between different users. It must always be possible to identify a responsible user for any observed action in the infrastructure.

Directive 2 (Delegation). *In case of vacation or illness the execution of activities must be delegated.*

In many organizations you have a clean desk policy, which means that at the end of the day all business relevant documents must be removed from the desk. Translating this into the digital world means that the desktop computer has to be switched off at the end of the day, which is also in the context of green IT getting more attention. This results in the following directive:

Directive 3 (Clean Desk Policy). *The user must shut down his/her computer at the end of the day, i.e., there are no open files or applications active anymore.*

Provenance relations are based on the observed reading and saving of files. Since it can not be observed by WebDAV when a file is closed again, the clean desk policy directive enforces that at the end of a day all files are closed. Thus, this is a synchronization point for deriving provenance relations by excluding files opened at previous days.

4.3 Provenance Relations

The generic directives are the basis to infer provenance relations based on WebDAV commands. A provenance relation is a relation between a read or save operation of a file

A and a save operation of a file B, where the read or save operation of file A is performed before the save operation of file B. The order of the operations can be determined by the timestamps at which the operations are observed and on the associated revision numbers.

Due to directive 3 only read operations which have been performed at the same day as the write operation are considered. Further, according to directive 2 each file access is associated with a specific user, thus, provenance relations require that the files are read and saved by the same user.

Based on the scenario described in Sect 3 it can be derived that e.g. the modification of a latex file will not use information from an image file. This independence, which is derivable from the process, has to be explicated as specific organizational directives as discussed in the following section.

4.4   Specific Organizational Directives

The goal of the organizational directives is to increase the precision of data provenance by excluding observed provenance relations which are actually irrelevant. Since provenance relations are based on sets of read files with a single point in time when files are certainly closed again (see Directive 3), additional mechanisms are required to determine the precise inference of data provenance. One way is to apply directives on the hierarchical structure of data on the network drive. Further such a structure is often established to exchange information between different organizations and to control access rights. In particular, one possible approach is to establish a basic directory structure where each directory is associated to a particular activity in a data processing workflow, as it is known e.g. from group ware solutions like BSCW or groove. An alternative is to use filters based on regular expressions exploiting file naming conventions or specific file extensions being unique for the output of an activity.

However, the derived directives might be not specific enough or too complicated to implement resulting in imprecise provenance information. The challenge is to find a good balance between usability of organizational directives and the targeted data provenance precision.

With regard to the motivating scenario in Sect 3 the following directives should be instantiated:

Directive 4 (Scientific paper writing)

1. *All bibtex files are located in a bib subdirectory of the project root directory.*
2. *All pictures and figures are located in a pics subdirectory of the project root directory.*
3. *All latex files, project files and auxiliary files are located in the project root directory.*
4. *Figures may require a conversion of file formats. The source figure filename is a prefix of the target figure filename ignoring file extensions.*
5. *An update of a bibtex file only depends on the old bibtex file, i.e. the filename of source and target are equivalent.*
6. *The execution of a pdflatex command reads data from all project directories and writes a pdf file in the project root directory.*

7. *The execution of a bibtex command reads from the project root directory and the bib directory and writes a bbl file in the project root directory.*

5   Provenance Relation Derivation

After the conceptual introduction in the previous section a more detailed technical discussion follows. The provenance information forms a provenance graph, which consists of vertices and edges. A vertex is either an access log entry or a SVN log entry. The edges represent the provenance relations. In this paper three classes of provenance relations are distinguished: provenance relation correlating SVN and access log entries, SVN step relations, and relations derived from directives. A access log contains read (GET), write (PUT), move (MOVE), and delete (DELETE) operations (WebDAV commands) applied to files. Further, the SVN internal logs contains write (Add, Modify), delete (Delete) and move (combined Add and Delete) entries.

All relation classes are discussed in the following subsections after a discussion on observations of file handling using WebDAV clients.

5.1   File Handling Observations

The manipulation of files via a WebDAV client and documenting them in SVN and access logs is not as straight forward as initially expected. In particular, the following observations can be made:

First, adding and updating files is realized in different combinations of WebDAV commands by different applications.

Second, the WebDAV MOVE command documents only the source filename of the move in the access log, but not the destination filename. However, by using WebDAV with autocommit each change on a file results in a new revision number, thus, only a MOVE command results in a delete and an add SVN log entry with the same revision number. This can be facilitated to correlate SVN and access log entries.

Third, a WebDAV DELETE command removes a file from the SVN repository. After the command is executed the file is not in the repository anymore, thus, the revision number of the removal of the file requires an extra query to the SVN log.

Fourth, reading a file is only documented in the access log. Which revision of a file has been read has to be inferred from the state of the SVN and the size of the file read.

The final observation is that the correlation between SVN and access log entries can not be based on time, since the timestamps recorded by the SVN and the access log are points in time when the event has been recorded in the corresponding system. Thus, the timestamp of the access log entry is always before the entry in the SVN log since the HTTP request is first processed by the HTTP server which forwards the request to the WebDAV and therefore to the SVN. However, in case of two fast subsequent operations, it is possible that the access log has two entries before any entry is recorded in the SVN log. Thus, it is sometimes hard to infer the correlation of access to SVN log entries.

Based on these observations, a simplified version of the correlation algorithm is depicted in Alg 1. The token variable indicates whether the SVN entry can be correlated to an access log entry. A correlation is possible if the SVN entry is either a Delete or an

Add entry (line 3). Further, a correlation is possible if the SVN entry is a Modify entry and the previous entry has been older than 2 seconds (line 4). The two second bound is based on the observation that an incremental upload of a file to the SVN resulted in a new revision approximately every second in our test system.

If an entry can not be correlated to an access log entry (line 7), than it is inferred that an incremental upload is occurring. Incremental updates are documented as a provenance relation between SVN log entries called *SVN step* relation (line 9). Otherwise, Add and Modify entries are correlated with PUT entries (line 11) and Delete entries are correlated either with DELETE (line 12) or with MOVE entries (line 13).

Since the file access is not documented in the SVN log entries, a second loop is executed on the access log entries (line 14). In particular, all GET entries are selected (line 15). For each GET entry the corresponding SVN entry is inferred (line 16) and the entries are correlated (line 17). In Fig 3 an example of the access and SVN log are

```
1  token=true;
2  forall the SVN entries do
3      if Delete or Add then token=true;
4      if Modify and size>0 and time difference to previous event<2sec then
5          token=true;
6      if not token then
7          ignore entry for correlation;
8          document SVN step relation with previous event;
9      else
10         if Add or Modify then correlate to PUT;
11         if Delete then
12             correlate to DELETE;
13             if not possible then correlate to MOVE;

14 forall the access log entries do
15     if GET then
16         find SVN entry preceding the GET with same filename and size;
17         correlate entries;
```

Algorithm 1. Simplified Data Fusion

depicted. Saving file *pics/prototype.vsd* (F1) results in a PUT entry in the access log and two entries in the SVN log: an Add entry and a Modify entry. The Add and Modify entry in the SVN log are related by a SVN step relation. Further, the PUT access log entry is correlated with the Add SVN log entry. Saving file *pics/prototype_architecture.eps* (F2) results in a single entry in the access and SVN log. Reading the file F1 results in a GET access log entry which is correlated to the last write operation of file F1 in the SVN log, i.e., the Modify entry resulting on version 2.

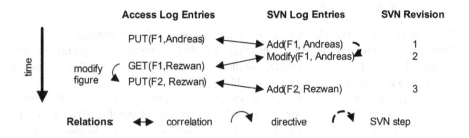

Fig. 3. Derived relations for the paper writing use case

5.2 Directive Provenance Relations

The last type of relations are the ones derived from organizational directives. This type of relation is inferred between access log entries only. As outlined in Sect 4 all files read at the same day as another file is saved are potentially contributing to the saved file. The relation is labeled by the associated activity in the process. The organizational directives can be translated into constraints on these potential provenance relations mainly by constraining files locations and filenames of sources and targets of a provenance relation.

For example Directives 4 in Sect 4.4 determine that only files in directory *pics* are relevant for the 'modify figure' activity (see Fig 1). Thus, files of other directories do not contribute to any provenance relation associated to this activity.

In Fig 3 an example of the 'modify figure' directive relation is depicted. Since file *pics/prototype.vsd* (F1) is a prefix (ignoring the extension) of file *pics/prototype_architecture.eps* (F2) as specified in directive 4.4 a directive relation between the *GET(F1, Rezwan)* access log entry and the *PUT(F2, Rezwan)* access log entry can be inferred.

6 Evaluation

To evaluate the proposed approach the running example described in Sect 3 is used. The scenario is a semi-structured process of writing a scientific paper using latex, where figures are usually created in Microsoft Visio, which are then converted to Enhanced Windows Metafile (emf) files, and further into Encapsulated PostScript (eps) files. The pdflatex command uses a library to convert the eps figure files into pdf figure files, which can then be used in the resulting pdf file.

The process uses the generic organizational directives (directives 1-3 in Sect 4.2) as well as the specific organizational directives (directives 4.1-7) discussed in Sect 4.4.

These organizational directives are not hard to implement, since we use the same way of structuring a paper writing project for years already. Our guess is that for many semi-structured processes this is similar, since people tend to organize their data rather on content and topics than on time.

6.1   Overhead

The paper is based on two latex files, which uses three style files contained in WebDAV. There are six bibtex files used and one image file with its corresponding emf, eps and pdf files. Further there is one BPMN file directly available as pdf file. The processing uses 4 auxiliary files and writes a single pdf file.

The overhead for the user during this experiment is clearly experienced due to the network delay introduced by using WebDAV compared to a local disc. My ADSL line at home is a bout a 1000 times slower than my local disc access. In particular, for a pdflatex command 840kB are read and 377kB are written, thus, in total 1.2MB are transferred. Further, for a bibtex command 250kB are read and 3kB are written, thus, in total 253MB are transferred. So far we have approximately 120 executions of the pdflatex command during the complete paper writing process. As a conclusion, since a build is not performed that often, the overhead is effecting our working experience only marginally.

The usage of WebDAV has an influence on the network load. How much depends on the size of the files the users are working with. In the context of latex these are rather small files and therefore the effects are hardly measurable. In case of other applications using e.g. spatial information or large image files with many write operations this may be different.

6.2   Quality

Quality assessment is problematic since there is no ground truth. Capturing the ground truth automatically is difficult. It would require to extend the operating system and monitor all content related actions, such as opening and closing files, copy and past operations from one application/file to another etc. This includes also the handling of content by non UI applications like e.g. running a pdflatex command. In this case file read system level routines have to be monitored. With the current available resources this is not feasible. In the provenance community there are groups of people researching on how to do this best, however, we are not aware of a complete solution yet - especially not for windows systems.

As a consequence the only possibility is to manually assess the quality of the inferred provenance information by inspection. The manual inspection does not show any missing provenance relations. Sometimes additional provenance relations are reported, which are e.g. artifacts of file transfers (temporary files). From our inspection we have not found cases with missing information. The precision depends on the organizational directives, which can be adjusted to the required level of precision as discussed before. Please be aware that the current alternative is no provenance information at all.

7   Mining Provenance Patterns

The mining approach uses the available log information and applies some counting and some statistic methods to infer provenance relations between different filenames. In particular, the assumption is if an activity is performed repeatedly then the same relations must show up in the log files. In a first step the time difference between file accesses is used to mine provenance relations and in a second step the frequency of file updates is considered.

7.1 Time Based Mining

Before the actual mining can start the available dataset has to be investigated with regards to the characteristics of accessing files. The time behavior of a system depends besides others on the used application, the system, and the network connection. Therefore it is not possible to provide an absolute number here.

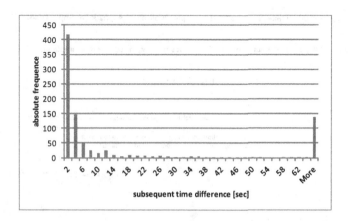

Fig. 4. Time difference in seconds between subsequent entries in the log file

Thus, a histogram of the time difference between two subsequent entries in the log is calculated (see Fig 4). In the histogram you can see that many files are accessed within a time difference of two seconds. The histogram turns flat for time differences greater than 12 seconds therefore, in the following we use the 12 seconds as an upper bound to infer provenance relations between files.

Next we retrieve all pairs of log entries which are less than 12 seconds apart and appear at least twice, remove transitive relations, and represent these relations as a weighted graph, where the weight corresponds to the number of observed occurrences of the relation. To enforce a directed weighted graph, the directionality is enforced by considering only the direction with the higher weight. The directed graphs are depicted in Fig 5 and 6, where the thickness of the connection corresponds to the weight of the relation.

The relations result actually in four unconnected graphs. The graph in Fig 6 corresponds to the provenance related to executing the *do pdflatex* activity and the *do bibtex* activity of the BPMN model (see Fig 1). The two activities are captured in one graph since the underlying dataset was produced by using a tool, which provides a macro to perform the sequence *pdflatex-pdflatex-bibtex-pdflatex* automatically. Therefore no sufficient big time difference can be observed between the provenance related to the *bibtex*

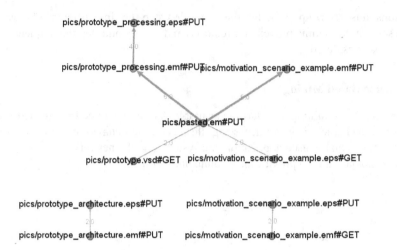

Fig. 5. Cluster 1

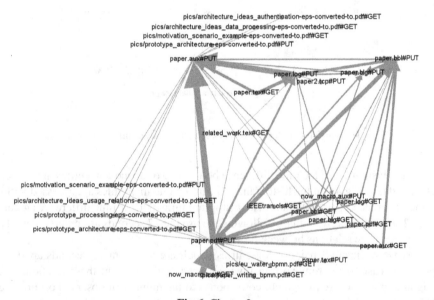

Fig. 6. Cluster 2

and the *pdflatex* activity. This graph also contains the conversion of figures from eps format to pdf format which is performed by a script related to the *pdflatex* command. This is visible by the writing of pdf files with the suffix *eps_converted-to*. The corresponding eps files are not included since the weights of the relations must be at least two.

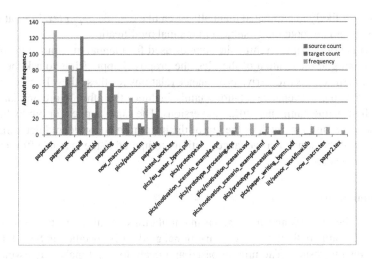

Fig. 7. Histogram of occurrence of files as a source or as a target in an identified provenance relation compared to the frequency of the file being written (incomplete)

An unexpected node in this graph is the writing of file *paper.tex* since it is not related to performing the two activities. The explanation for this is that the used tool before executing the above sequence of *bibtex* and *pdflatex* commands checks whether all source files have been saved. If this is not the case then the tool automatically saves the modified files. Therefore, the writing of the latex source file *paper.tex* becomes a part of this graph.

The graphs in Fig 5 are related to figure conversion. It contains the conversion of figures from the emf format into eps format initiated by the author using a conversion tool. The conversion of the other figures is not represented in this graph since a frequency of at least two relations was required. The graph containing the visio file (vsd extension) represents the copying of a visio figure into an emf file.

Overall it can be concluded that all graphs are related to some automated processing initiated by the user. However, manual modifications of files are not observed since they usually require more than 12 seconds to perform the modification, like e.g. writing a new section in the paper.

7.2 Frequency Analysis

Next we try to identify the files which are mainly manipulated manually. To do so we calculate the histogram of files being written as partially depicted in Fig 7. This histogram is compared with the histogram of the files being part of the provenance relations as discussed in the previous subsection. A file can be a source or the target of the provenance relation. In Fig 7 the two ways of participation are explicated per file. If a file is written often and participates a lot in provenance relations like e.g. file *paper.pdf*, than it is unlikely that this file is subject to manual manipulation. On the other hand side if a file hardly participates in a provenance relation but is written often then it is most likely manipulated manually often like e.g. file *paper.tex*. Files at the tail of the histogram are potentially subject to manual modification.

Since it is not possible to mine the sources used for manual manipulations of a file without looking at the content of the files, the proposed approach is to ask the user to provide patterns on how to derive provenance relations for these files. In case of figure and latex files the provenance relation mainly includes the other figure and latex files often with the same name, just different versions. The same applies for bibtex files.

The decision on how much of the tail of the histogram is considered determines the imprecision of the derived provenance relations compared to the manual effort required to explicate these rules.

7.3   Evaluation

Comparing the clusters and the manual manipulations identified in the previous two sections and compare them to the organizational guidelines specified in Sect 4.4 shows some interesting results: The mining based approach does not make any assumptions on the behavior of the user. However, in case of manually modified files user input is required to derive the provenance relations. The files to which this applies can be mined automatically.

The organizational directions distinguish the *pdflatex* and *bibtex* activity. However, the mining approach identified additional automated activities like the conversion of emf to eps files and the conversion from eps to pdf files. This indicates that the organizational guidelines may be incomplete.

Comparing the mined graph for *bibtex* and *pdflatex* activity with the provenance derived using organizational guidelines it turns out that the mined approach covers pretty well the real provenance although it may have some additional provenance relations. This is because the processing of *bibtex* and *pdflatex* command in a sequence by the tool results in mined provenance relations which are actually not existent.

8   Conclusion

In this paper, a provenance capturing approach for semi-structured processes involving provenance unaware legacy systems is proposed. Further, the derivation of different classes of provenance relations is discussed. It has been argued that the recall of the proposed approach is very high while the precision depends on used organizational directives, i.e., constraints on handling files as a basis for deriving provenance relations.

Future work should address the currently used 'version on every write' approach to minimize the versions used. A further topic is instead of deriving organizational directives to observe directives and assess the quality of the data handling as applied by the users for deriving provenance relations.

References

1. Ludascher, B., Altintas, I., Berkley, C., Higgins, D., Jaeger, E., Jones, M., Lee, E., Tao, J., Zhao, Y.: Scientific workflow management and the Kepler system. Concurrency and Computation: Practice and Experience 18(10), 1039–1065 (2006)

2. Oinn, T., Addis, M., Ferris, J., Marvin, D., Greenwood, M., Carver, T., Pocock, M., Wipat, A., Li, P.: Taverna: a tool for the composition and enactment of bioinformatics workflows. Bioinformatics 20(17), 3045–3054 (2004)

3. Huq, M.R., Wombacher, A., Apers, P.M.G.: Facilitating fine grained data provenance using temporal data model. In: Proc. 7. Intl Workshop on Data Management for Sensor Networks, DMSN, pp. 8–13. ACM (September 2010)

4. Cui, Y., Widom, J.: Lineage tracing for general data warehouse transformations. VLDB Journal 12(1), 41–58 (2003)

5. Szomszor, M., Moreau, L.: Recording and reasoning over data provenance in web and grid services. In: Meersman, R., Schmidt, D.C. (eds.) CoopIS 2003, DOA 2003, and ODBASE 2003. LNCS, vol. 2888, pp. 603–620. Springer, Heidelberg (2003)

6. Simmhan, Y.L., Plale, B., Gannon, D.: A survey of data provenance in e-science. SIGMOD Rec. 34(3), 31–36 (2005)

7. Allen, M.D., Chapman, A., Blaustein, B., Seligman, L.: Capturing provenance in the wild. In: McGuinness, D.L., Michaelis, J.R., Moreau, L. (eds.) IPAW 2010. LNCS, vol. 6378, pp. 98–101. Springer, Heidelberg (2010)

8. Seltzer, M., Muniswamy-Reddy, K.K., Holland, D.A., Braun, U., Ledlie, J.: Provenance-aware storage systems. In: Proceedings of the USENIX Annual Technical Conference, USENIX 2006 (June 2006)

9. Margo, D.W., Seltzer, M.I.: The case for browser provenance. In: Cheney, J. (ed.) Workshop on the Theory and Practice of Provenance. USENIX (2009)

10. Futrelle, J.: Tupelo server, http://tupeloproject.ncsa.uiuc.edu/

11. Simmhan, Y.L., Plale, B., Gannon, D.: Karma2: Provenance management for data driven workflows. Intl. J. of Web Services Research 5, 1–23 (2008)

12. Moreau, L., Freire, J., Futrelle, J., McGrath, R.E., Myers, J., Paulson, P.: The open provenance model: An overview. In: Freire, J., Koop, D., Moreau, L. (eds.) IPAW 2008. LNCS, vol. 5272, pp. 323–326. Springer, Heidelberg (2008)

13. Misra, A., Blount, M.L., Kementsietsidis, A., Sow, D., Wang, M.: Advances and Challenges for Scalable Provenance in Stream Processing Systems. In: Freire, J., Koop, D., Moreau, L. (eds.) IPAW 2008. LNCS, vol. 5272, pp. 253–265. Springer, Heidelberg (2008)

Managing Structural and Textual Quality
of Business Process Models

Jan Mendling

Wirtschaftsuniversität Wien, Augasse 2-6, A-1090 Vienna, Austria
jan.mendling@wu.ac.at

Abstract. Business process models are increasingly used for capturing business operations of companies. Such models play an important role in the requirements elicitation phase of to-be-created information systems and in as-is analysis of business efficiency. Many process modeling initiatives have grown considerably big in size involving dozens of modelers with varying expertise creating and maintaining hundreds, sometimes thousands of models. One of the roadblocks towards a more effective usage of these process models is the often insufficient provision of quality assurance. The aim of this paper is to give an overview on how empirical research informs structural and textual quality assurance of process models. We present selected findings and show how they can be utilized as a foundation for novel automatic analysis techniques.

1   Introduction

Nowadays, many companies document their business processes in terms of conceptual models. These models provide the basis for activities associated with the business process management lifecycle such as process analysis, process redesign, workflow implementation and process evaluation. Many process modeling initiatives have resulted in hundreds or thousands of process models created by process modelers of diverging expertise. One of the major roadblocks towards a more effective usage of these process models is the often insufficient provision of quality assurance. This observation establishes the background for the definition of automatic analysis techniques, which are able to support quality assurance.

In recent years, research into quality assurance of process models and corresponding analysis techniques has offered various new insights. The objective of this paper is to summarize some of the essential contributions in this area. To this end, we aim to integrate both technical contributions and empirical findings. The paper is structured accordingly. In Section 2 we describe the background of quality research distinguishing structural and textual quality. Section 3 discusses how quality factors can be analyzed in terms of their capability to predict aspects of quality. Section 4 discusses different techniques for automatically refactoring process models with the aim to improve their quality. Finally, Section 5 summarizes the discussion and concludes the paper.

P. Cudre-Mauroux, P. Ceravolo, and D. Gašević (Eds.): SIMPDA 2012, LNBIP 162, pp. 100–111, 2013.

2   Background

Research on conceptual modeling often distinguishes syntax, semantics and pragmatics of process models with a reference to semiotic theory [1,2]. The idea behind this distinction is that a message, here codified as a conceptual model, first has to be understood in terms of its syntax by a model reader before the semantics can be interpreted. Comprehension on the semantic level then provides the foundation for taking appropriate action in a pragmatic way. This semiotic ladder has one major implication for process modeling as a specific area of conceptual modeling and one major research directive. The implication of a semiotic perspective on process modeling is that the comprehension of a process model by a model reader has to be regarded as the central foundation for discussing its quality. As appropriate pragmatics, which comes down to actions taken by a model reader, define the successful progression on the semiotic ladder, research has to establish a thorough understanding how quality on each step of this ladder can be achieved. Indeed, it has been shown empirically that none of the three steps of the semiotic ladder can be neglected, and that all steps appear to be of equal importance for conceptual modeling [3]. As much of research on process modeling has advanced analysis of syntax and execution semantics of process models, but rather neglected textual semantic and pragmatic aspects, it is an important directive for future research to complement syntactic analyses with insights on semantics and pragmatics. In the following, we try to give a balanced account of research on process model quality on a syntactic and semantic level while leaving out pragmatics. Our focus in this context is on structural and textual characteristics of a process model.

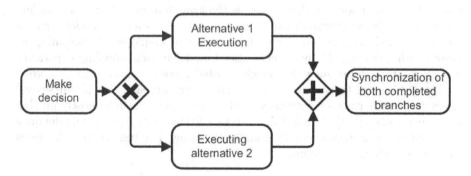

Fig. 1. Example of a process model with structural and textual issues

Figure 1 shows the example of a simple process model in BPMN notation. Process models like this one define the temporal and logical constraints on the control flow between different activities of the process. Here, there are four activities: *Make decision*, *Alternative 1 Execution*, *Executing alternative 2*, and *Synchronization of both completed branches*. The textual labels of these activities describe on the level of domain semantics what this process is supposed to do. The activity nodes together with the gateways and arcs define the syntax or the formal structure of the process model. In this model, there are two types of gateways used. The first one, an XOR-split, defines a decision point to progress either with the upper or the lower branch, but never with both. There is also a corresponding XOR-join in BPMN, it is not used in the example. Towards the right-hand side of the model, there is an AND-join. This element is used to synchronize concurrent branches. There is no corresponding AND-split in the model. The arcs define the flow relation between activity nodes and gateways.

The quality of a process model like the one in the example can be discussed from the perspective of syntax and of semantics. The quality of the syntax of the model relates to the question whether its formal structure can be readily understood by a model reader. In this context, prior research has focused on the question whether the size and the complexity might be overwhelming. Furthermore, there are formal correctness criteria that can be automatically checked. For the example, we can see that it apparently includes a deadlock: the single branch activated by the XOR-split eventually activates the AND-join, which will then wait forever for the not activated alternative branch to complete. The quality of the semantics of the model relates to the question whether its textual content can be readily understood by a model reader. Here, we observe that the activity labels follow different grammatical structure. *Make decision* starts with a verb and continues with a business object. This is usually considered to be the norm structure of an activity label [4,5,6,7,8]. The other three labels use a gerund or a noun to express the work content of the activity. Altogether, we can summarize that the example model has both issues with its syntax and with its semantics.

In practice a considerable percentage of process models has quality issues, with often 5% to 30% of the models having problems with soundness [9]. The reason for at least some of these issues is the growth of many process modeling initiatives. This development causes problems at the stage of model creation and model maintenance. An increasing number of employees is becoming involved with modeling. Many of these casual modelers lack modeling experience and adequate training such that newly created models are not always of good quality [10,11]. Furthermore, the fact that many companies maintain several thousand models calls for automatic quality assurance, which is mostly missing in present tools [10,11]. A promising direction for increasing process model quality is automatic guideline checking and refactoring. The next section discusses the corresponding foundations.

3   Factors of Process Model Understanding

Various factors for process model understanding have been identified. Characteristics of the *modeling notation* have be investigated in several experiments [12,13,14]. Two different factors have to be discussed in this context. First, ontological problems of the notation, e.g. when there are two options to represent the same meaning, might lead to misinterpretations of singular models [15]. Survey research has found support for this argument [16]. Second, properties of the symbol set of a notation might cause problems, e.g. with remembering or distinguishing them [17].

Empirical support for this hypothesis is reported in [18]. The *secondary notation* plays an important role as well. The concept of secondary notation covers all representational aspects of a model that do not relate to its formal structure. This might relate to the usage of color as a means of highlighting [19]. Corresponding support was found in an experiment in [20]. The visual layout of the model graph is also well-known for its importance to facilitate good understanding [21,22]. In this section, we focus on structural properties of the process model and properties of its text labels.

3.1 Structural Factors of Process Model Understanding

Structural properties of a process model are typically operationalized by the help of different metrics. Many of them are inspired by metrics from software engineering like lines of code, cyclomatic number, or object metrics [23,24,25]. Early contributions in the field of process modeling focus on the definition of metrics [26,27,28]. More recent work puts a strong emphasis on the validation of metrics. In these works, different sets of metrics are used as input variables for conducting experiments to test their statistical connection with dependent variables that relate to quality. For instance, the control-flow complexity (CFC) [29] is validated with respect to its correlation with perceived complexity of models [30]. Metrics including size, complexity and coupling are validated for their correlation with understanding and maintainability [31,32]. Further metrics aim to quantify cognitive complexity and modularity [33,34,35,36]. Various metrics have been validated as predictors of error probability [37], which is assumed to be a symptom of bad understanding by the modeler during the process of model creation. A summary of metrics is presented in [38], an overview of experiments can be found in [39,40]. In summary, it can be stated that increase in size, decrease in complexity and decrease in structuredness leads is related to greater issues with quality.

One of the major objectives of research into the factors of process model understanding is to establish a set of sound and precise guidelines for process modeling. Guidelines such as the Guidelines of Process Modeling [41] have been available for a while, but they had hardly been tied to experimental findings. The Seven Process Modeling Guidelines (7PMG) might be regarded as a first attempt towards building guidelines based on empirical insight [6]. The challenge in this context is to adapt statistical methods in such a way that metrics can be related to threshold values. In its most basic form, this problem can be formulated as a classification problem: if we consider a particular metric like number of nodes, in how far is it capable of distinguishing e.g. good and bad models.

A specific stream of research in this area investigates in how far different process model metrics are capable of separating models with and without errors. The work reported in [42] uses logistic regression and error probability as a dependent variable. Logistic regression is a statistical model for estimating the probability of binary choices (error or no error in this case) [43]. The logistic regression estimates the odds of error or no error based on the logit function. This model can be adapted by using structural metrics such as size or complexity of a process model as input variables and observations in terms of whether these models are sound or not. The relationship between input and dependent variables follows an S-shaped curve of the logit curve converging to 0 for $-\infty$ and to 1 for ∞. The value 0.5 is used as a cut-off for predicting error or no error. Based on the coefficient of the input variables in the logit function, one can predict whether an error would be in the model or not.

The quality of such a function to classify process models correctly as having an error or not can be judged based on four different sets: the set of true positive (TP) classifications, the set of false positives (FP), the set of true negatives (TN) and the set of false negatives (FN). A perfect classification based on the logit function would imply that there are only true positives and true negatives. An optimal threshold of separation can then be determined using Receiver Operating Characteristic (ROC) curves [44]. These curves visualize the ability of a specific process metric to discriminate between error and no error models. Each point on the ROC curve defines a pair of *sensitivity* and $1 - specificity$ values of a metric. The best threshold can then be found based on sensitivity and specificity values with: $sensitivity = $ true positives(TP) rate $= \frac{TP}{P}$, $specificity = 1 - $ false positives(FP) rate $= \frac{1-FP}{P}$. Using this approach, several guidelines of the 7PMG could be refined in [42]. Table 1 provides an overview of the results showing, among others, that process models with more than 30 nodes should be decomposed.

Table 1. Ten Process Modeling Rules

Rule	Associated measure	Explanation
G1	Nodes	Do not use more than 31.
G2	Conn. Degree	No more than 3 inputs or outputs per connector.
G3	Start and End	Use no more than 2 start and end events.
G4.a	Structuredness	Model as structured as possible.
G4.b	Mismatch	Use design patterns to avoid mismatch.
G5.a	OR-connectors	Avoid OR-joins and OR-splits.
G5.b	Heterogeneity	Minimize the heterogeneity of connector types.
G5.c	Token Split	Minimize the level of concurrency.
G6	Text	Use verb-object activity labels.
G7	Nodes	Decompose a model with more than 31 nodes.

Although there have been considerable advancements in this area, there are several challenges that persist. Thresholds have been identified based on error probability as a dependent variable, which can be easily expressed in a binary way. An important antecedent of quality is understanding. However, thresholds of understanding are much more difficult to establish as it is mostly measured using score values summed up for a set of comprehension tasks. In this case, good and bad models cannot be exactly discriminated. Furthermore, understanding can be associated with different types of comprehension questions ranging from simple

Table 2. Activity Labeling Styles

Labeling Style	Structure	Example
Verb-Object	Action(Infinitive) + Object	Submit Letter
Action-Noun (np)	Object + Action(Noun)	Letter Submission
Action-Noun (of)	Action(Noun) + 'of' + Object	Submission of Letter
Action-Noun (gerund)	Action(Gerund) + Object	Submitting Letter
Action-Noun (irregular)	*undefined*	Submission: Letter
Descriptive DES	Role + Action(3P) + Object	Student submits Letter
No Action	*undefined*	Letter

recall of a model, understanding its semantics to pragmatic problem solving tasks. Up until now, it has not been studied in how far the same or different metrics influence each of these comprehension tasks.

3.2 Labeling Style as a Factor of Process Model Understanding

Empirical research has found that process models from practice do not always follow naming conventions such as the verb-object style for activities. There are three general classes of activity labeling styles [4] (see Figure 2). First, the verb-object style defines an activity label as a verb followed by the corresponding business object (*Make decision*). Second, there are different ways of defining activities as action-noun labels. For such a label, the action is not formulated as a verb, but rather as a gerund (*Executing*) or a substantivated verb (*Execution* from *to execute*). There is also a third category of activity labels that miss referring to an action. An example is the label *information system*, which fails to mention an action, neither as a verb or noun.

With these categories defined, it has to be noted that labeling style is a factor with characteristics quite different to structural metrics. While metrics can be measured on a metric scale, labeling styles can only be distinguished in a nominal way. This means that in the simplest case the input variable can be defined in a binary way, distinguishing usage of verb-object style versus usage of another style. In terms of defining quality preferences, this makes the task easier: while metrics require a threshold to distinguish good and bad, labeling styles can be directly compared to be better or worse. An experiment reported in [4] takes activity labels of different labeling styles as treatments in order to investigate their potential ambiguity and their usefulness in facilitating domain understanding. ANOVA tests demonstrate that verb-object labels are perceived to be significantly better in this regard, followed by action-noun labels. Labels of the rest category were judged to be most ambiguous.

While the usage of labeling style is covered well in the literature, there are still various challenges in dealing with terminology. From a quality perspective, terms should have a clear-cut meaning. This implies that synonyms (several words with the same meaning) and homonyms (several meanings of the same word) should be avoided in process modeling. This problem is acknowledged in various papers [45,46,47]; however, a proper solution for automatic quality assurance is missing.

4   Automatic Refactoring

The empirical results reported in the previous section provide a basis for the development of automatic refactoring techniques. The general idea of refactoring was formulated for software and relates to "changing a software system in such a way that it does not alter the external behavior of the code, yet improves its internal structure" [48]. For process models, often the notion of trace equivalence [49] or one of the notions of bisimulation [50] is considered when refactoring models. In the following, we summarize work on refactoring the structure of a process model and its activity labels. Frameworks for categorizing refactorings have been proposed in [49,19,51].

4.1   Refactoring the Structure of Process Models

Insight into factors of process model comprehension provides a solid basis for optimizing its structure. The challenge in this context is to define a transformation from an unstructured model towards a structured model. It is well known that a structured model can always be constructed for process models without concurrency, but that some concurrent behaviour is inherently unstructured [52]. The research reported in [50] presents a approach based on the identification of ordering relations which leads to a maximally structured model under fully concurrent bisimulation.

Here, two cases have to be distinguished. There are process models for which making them structured comes at the price of increasing its size. Such a case is shown in Figure 2. This increase stems from the duplication of activities in unstructured paths. There are also cases where a process model can be structured without having to duplicate activities. In practice, making a model structured without duplication appears to be rather rare. An investigation with more than 500 models from practice has shown that structuring leads to an increase in size of about 50% on average [53]. It is also important to note that duplication might be more harmful than a usual increase in size. The user experiment reported in [53] points to a potential confusion by model readers who are asked about behavioural constraints that involve activities that are shown multiple times in the model.

The problem of duplicating activities is a key challenge in this area. It is an open research question how the beneficial effects of structuring can be best balanced with the harmful introduction of duplicate activities. Further experiments are needed for identifying a precise specification of the trade-off between structuredness and duplication. In this context, also the size of the model has to be taken into account.

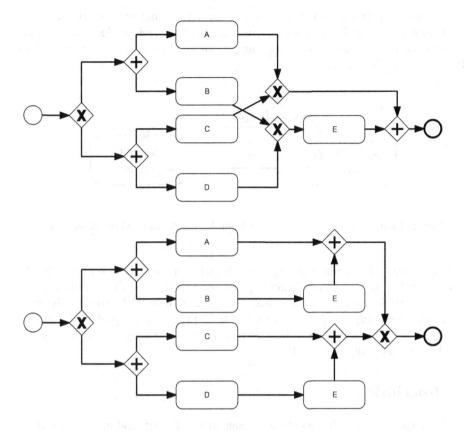

Fig. 2. Example of an unstructured and corresponding structured process model

4.2   Refactoring Text Labels of Process Models

Experiments and best practices from industry suggest a preference for the verb-object labeling style. The challenge in this context is to achieve an accurate parsing of the different labeling styles such that they can be transformed to the verb-object style. An accurate parsing is difficult in English for two reasons. First, many nouns in English are built from a verb using a zero-derivation mechanism. This means that the noun is morphologically equivalent to the verb. For a word like *plan* we do not directly know whether it refers to a verb or a noun (*the plan* versus *to plan*). Second, the activity labels of a process model usually do not cover a complete grammatically correct sentence structure. Therefore, it has been found difficult to use standard natural language processing tools like the Stanford parser. The approach reported in [54] uses different contextual information to map a label that, for instance, starts with the word *plan* to its correct labeling style. Once the labeling style is known, tools like WordNet can be used to find a verb that matches an action that was formulated as a noun (see Figure 3). It has been shown that this approach works accurately for three different modeling collections from practice including altogether more than 10,000 activity labels [54].

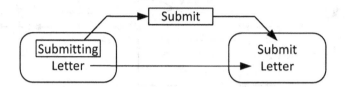

Fig. 3. Example of a label refactored from Action-Noun to Verb-Object style

It is a topic of ongoing research how these refactoring techniques can be defined in such a way that they do not depend upon the rich set of natural language processing tools available for English. An alternative could be to directly work with annotated corpora. Also, and related to the terminology problem identified above, it is up until now not clear how problems of synonyms and homonyms can be automatically reworked.

5   Conclusion

In this paper we have discussed the management of structural and textual quality of business process models. The growth of many process modeling initiatives towards involving dozens of modelers with varying expertise creating and maintaining thousands of models raises the question of how quality assurance can be defined and implemented in an automatical way. Insights into the factors of process model understanding provide the foundation for building such automatic techniques. On the structural side of process model quality, size and structuredness have been found to be major factors. Guidelines like 7PMG have been formulated based on empirical findings, pointing to the need for rework when certain thresholds are surpassed.

A topic of ongoing research is how refactoring techniques can be defined that balance different structural properties such as size and structuredness while minimizing the duplication of activities. On the side of activity labels, the usage of the verb-object style is recommended. Automatic techniques in this context have to provide an accurate parsing of the labels with a potential reformulation of actions that might be stated as nouns. In this area it is a topic of ongoing research to what extent such automatic techniques for style recognition can be defined without relying on tools like WordNet such that they can be adapted for languages different to English.

References

1. Lindland, O., Sindre, G., Sølvberg, A.: Understanding quality in conceptual modeling. IEEE Software 11(2), 42–49 (1994)
2. Krogstie, J., Sindre, G., Jørgensen, H.: Process Models Representing Knowledge for Action: a Revised Quality Framework. European Journal of Information Systems 15(1), 91–102 (2006)
3. Moody, D.L., Sindre, G., Brasethvik, T., Sølvberg, A.: Evaluating the quality of process models: Empirical testing of a quality framework. In: Spaccapietra, S., March, S.T., Kambayashi, Y. (eds.) ER 2002. LNCS, vol. 2503, pp. 380–396. Springer, Heidelberg (2002)
4. Mendling, J., Reijers, H.A., Recker, J.: Activity Labeling in Process Modeling: Empirical Insights and Recommendations. Information Systems 35(4), 467–482 (2010)
5. Silver, B.: BPMN Method and Style, with BPMN Implementer's Guide, 2nd edn. Cody-Cassidy Press (January 2011)
6. Mendling, J., Reijers, H.A., van der Aalst, W.M.P.: Seven Process Modeling Guidelines (7PMG). Information and Software Technology 52(2), 127–136 (2010)
7. Allweyer, T.: BPMN 2.0 - Business Process Model and Notation, 2nd edn. Books on Demand GMBH, Norderstedt (2009)
8. Leopold, H., Smirnov, S., Mendling, J.: On the refactoring of activity labels in business process models. Information Systems 37(5), 443–459 (2012)
9. Mendling, J.: Empirical Studies in Process Model Verification. In: Jensen, K., van der Aalst, W.M.P. (eds.) Transactions on Petri Nets and Other Models of Concurrency II. LNCS, vol. 5460, pp. 208–224. Springer, Heidelberg (2009); Special Issue on Concurrency in Process-Aware Information Systems 2
10. Rosemann, M.: Potential Pitfalls of Process Modeling: Part A. Business Process Management Journal 12(2), 249–254 (2006)
11. Rosemann, M.: Potential pitfalls of process modeling: part b. Business Process Management Journal 12(3), 377–384 (2006)
12. Sarshar, K., Loos, P.: Comparing the control-flow of EPC and petri net from the end-user perspective. In: van der Aalst, W.M.P., Benatallah, B., Casati, F., Curbera, F. (eds.) BPM 2005. LNCS, vol. 3649, pp. 434–439. Springer, Heidelberg (2005)
13. Hahn, J., Kim, J.: Why are some diagrams easier to work with? effects of diagrammatic representation on the cognitive integration process of systems analysis and design. ACMTCHI: ACM Transactions on Computer-Human Interaction 6 (1999)
14. Agarwal, R., De, P., Sinha, A.: Comprehending object and process models: An empirical study. IEEE Transactions on Software Engineering 25(4), 541–556 (1999)
15. Weber, R.: Ontological Foundations of Information Systems. Coopers & Lybrand and the Accounting Association of Australia and New Zealand, Melbourne, Australia (1997)
16. Recker, J., Rosemann, M., Green, P.F., Indulska, M.: Do ontological deficiencies in modeling grammars matter? MIS Quarterly 35(1), 57–79 (2011)
17. Moody, D.L.: The "physics" of notations: Toward a scientific basis for constructing visual notations in software engineering. IEEE Trans. Software Eng. 35(6), 756–779 (2009)
18. Figl, K., Mendling, J., Strembeck, M.: The influence of notational deficiencies on process model comprehension. Journal of the Association for Information Systems (2012) (in press)

19. Rosa, M.L., ter Hofstede, A.H.M., Wohed, P., Reijers, H.A., Mendling, J., van der Aalst, W.M.P.: Managing process model complexity via concrete syntax modifications. IEEE Trans. Industrial Informatics 7(2), 255–265 (2011)
20. Reijers, H.A., Freytag, T., Mendling, J., Eckleder, A.: Syntax highlighting in business process models. Decision Support Systems 51(3), 339–349 (2011)
21. Moher, T., Mak, D., Blumenthal, B., Leventhal, L.: Comparing the Comprehensibility of Textual and Graphical Programs: The Case of Petri Nets. In: Cook, C., Scholtz, J., Spohrer, J. (eds.) Empirical Studies of Programmers: Fifth Workshop: Papers Presented at the Fifth Workshop on Empirical Studies of Programmers, December 3-5, pp. 137–161. Ablex Pub. (1993)
22. Purchase, H.: Which aesthetic has the greatest effect on human understanding? In: DiBattista, G. (ed.) GD 1997. LNCS, vol. 1353, pp. 248–261. Springer, Heidelberg (1997)
23. McCabe, T.: A complexity measure. IEEE Transactions on Software Engineering 2(4), 308–320 (1976)
24. Chidamber, S., Kemerer, C.: A metrics suite for object oriented design. IEEE Transaction on Software Engineering 20(6), 476–493 (1994)
25. Fenton, N., Pfleeger, S.: Software Metrics. A Rigorous and Practical Approach. PWS, Boston (1997)
26. Lee, G., Yoon, J.M.: An empirical study on the complexity metrics of petri nets. Microelectronics and Reliability 32(3), 323–329 (1992)
27. Nissen, M.: Redesigning reengineering through measurement-driven inference. MIS Quarterly 22(4), 509–534 (1998)
28. Morasca, S.: Measuring attributes of concurrent software specifications in petri nets. In: METRICS 1999: Proceedings of the 6th International Symposium on Software Metrics, pp. 100–110. IEEE Computer Society, Washington, DC (1999)
29. Cardoso, J.: Evaluating Workflows and Web Process Complexity. In: Workflow Handbook 2005, pp. 284–290. Future Strategies, Inc., Lighthouse Point (2005)
30. Cardoso, J.: Process control-flow complexity metric: An empirical validation. In: Proceedings of IEEE International Conference on Services Computing, IEEE SCC 2006, Chicago, USA, September 18-22, pp. 167–173. IEEE Computer Society (2006)
31. Canfora, G., García, F., Piattini, M., Ruiz, F., Visaggio, C.: A family of experiments to validate metrics for software process models. Journal of Systems and Software 77(2), 113–129 (2005)
32. Aguilar, E.R., García, F., Ruiz, F., Piattini, M.: An exploratory experiment to validate measures for business process models. In: First International Conference on Research Challenges in Information Science, RCIS (2007)
33. Vanderfeesten, I., Reijers, H.A., Mendling, J., van der Aalst, W.M.P., Cardoso, J.: On a Quest for Good Process Models: The Cross-Connectivity Metric. In: Bellahsène, Z., Léonard, M. (eds.) CAiSE 2008. LNCS, vol. 5074, pp. 480–494. Springer, Heidelberg (2008)
34. Vanhatalo, J., Völzer, H., Leymann, F.: Faster and more focused control-flow analysis for business process models through SESE decomposition. In: Krämer, B.J., Lin, K.-J., Narasimhan, P. (eds.) ICSOC 2007. LNCS, vol. 4749, pp. 43–55. Springer, Heidelberg (2007)
35. van der Aalst, W.M.P., Lassen, K.: Translating unstructured workflow processes to readable BPEL: Theory and implementation. Information and Software Technology 50(3), 131–159 (2008)
36. Reijers, H.A., Mendling, J., Dijkman, R.M.: Human and automatic modularizations of process models to enhance their comprehension. Inf. Syst. 36(5), 881–897 (2011)

37. Mendling, J., Verbeek, H.M.W., Dongen, B., van der Aalst, W.M.P., Neumann, G.: Detection and Prediction of Errors in EPCs of the SAP Reference Model. Data & Knowledge Engineering 64(1), 312–329 (2008)
38. Mendling, J.: Metrics for Process Models. LNBIP, vol. 6. Springer, Heidelberg (2008)
39. Reijers, H.A., Mendling, J.: A Study Into the Factors That Influence the Understandability of Business Process Models. IEEE Transactions on Systems, Man, and Cybernetics, Part A 41(3), 449–462 (2011)
40. Mendling, J., Strembeck, M., Recker, J.: Factors of process model comprehension - findings from a series of experiments. Decision Support Systems 53(1), 195–206 (2012)
41. Becker, J., Rosemann, M., von Uthmann, C.: Guidelines of Business Process Modeling. In: van der Aalst, W.M.P., Desel, J., Oberweis, A. (eds.) Business Process Management. LNCS, vol. 1806, pp. 30–49. Springer, Heidelberg (2000)
42. Mendling, J., Sánchez-González, L., García, F., Rosa, M.L.: Thresholds for error probability measures of business process models. Journal of Systems and Software 85(5), 1188–1197 (2012)
43. Hosmer, D., Lemeshow, S.: Applied Logistic Regression, 2nd edn. John Wiley & Sons (2000)
44. Zweig, M., Campbell, G.: Receiver-operating characteristic (roc) plots: a fundamental evaluation tool in clinical medicine. Clinical Chemistry 39(4), 561–577 (1993)
45. Dean, D., Lee, J., Orwig, R., Vogel, D.: Technological support for group process modeling. Journal of Management Information Systems, 43–63 (1994)
46. Rosemann, M., Muehlen, M.: Evaluation of workflow management systems-a meta model approach. Australian Journal of Information Systems 6, 103–116 (1998)
47. Rolland, C.: L'e-lyee: coupling l'ecritoire and lyeeall. Information & Software Technology 44(3), 185–194 (2002)
48. Fowler, M., Beck, K., Brant, J., Opdyke, W., Roberts, D.: Refactoring: improving the design of existing code. Addison-Wesley Professional (1999)
49. Weber, B., Reichert, M., Mendling, J., Reijers, H.A.: Refactoring large process model repositories. Computers in Industry 62(5), 467–486 (2011)
50. Polyvyanyy, A., García-Bañuelos, L., Dumas, M.: Structuring acyclic process models. Inf. Syst. 37(6), 518–538 (2012)
51. Rosa, M.L., Wohed, P., Mendling, J., ter Hofstede, A.H.M., Reijers, H.A., van der Aalst, W.M.P.: Managing process model complexity via abstract syntax modifications. IEEE Trans. Industrial Informatics 7(4), 614–629 (2011)
52. Kiepuszewski, B., ter Hofstede, A.H.M., Bussler, C.J.: On structured workflow modelling. In: Wangler, B., Bergman, L.D. (eds.) CAiSE 2000. LNCS, vol. 1789, pp. 431–445. Springer, Heidelberg (2000)
53. Dumas, M., La Rosa, M., Mendling, J., Mäesalu, R., Reijers, H.A., Semenenko, N.: Understanding business process models: The costs and benefits of structuredness. In: Ralyté, J., Franch, X., Brinkkemper, S., Wrycza, S. (eds.) CAiSE 2012. LNCS, vol. 7328, pp. 31–46. Springer, Heidelberg (2012)
54. Leopold, H., Smirnov, S., Mendling, J.: On the refactoring of activity labels in business process models. Inf. Syst. 37(5), 443–459 (2012)

Author Index